CAMBRIDGE GEOGRAPHICAL STUDIES
Editorial Board: B. H. FARMER, A. T. GROVE, E. A. WRIGLEY

4 · FREIGHT FLOWS AND SPATIAL
ASPECTS OF THE BRITISH ECONOMY

CAMBRIDGE GEOGRAPHICAL STUDIES

FREIGHT FLOWS AND SPATIAL ASPECTS OF THE BRITISH ECONOMY

by MICHAEL CHISHOLM

Professor of Economic and Social Geography, University of Bristol

and PATRICK O'SULLIVAN

Assistant Professor of Geography, Northwestern University

CAMBRIDGE

AT THE UNIVERSITY PRESS 1973

Published by the Syndics of the Cambridge University Press
Bentley House, 200 Euston Road, London NW1 2DB
American Branch: 32 East 57th Street, New York, N.Y. 10022

© Cambridge University Press 1973

Library of Congress Catalogue Card Number: 72 835 92

ISBN: 0 521 08672 8

Text set in 10pt. IBM Press Roman, printed by photolithography,
and bound in Great Britain at The Pitman Press, Bath

3-1303-00060-4216

CONTENTS

LIST OF FIGURES

LIST OF TABLES

PREFACE

In 1965, Mr (now Professor) C. D. Foster drew Michael Chisholm's attention to the existence in the then Ministry of Transport of road traffic data for 1962 on the basis of 107 geographical zones. Through the good offices of Professor M. E. Beesley, he was able to obtain a copy of the matrix of total flows and in the ensuing year or two carried out some conceptually simple experiments to the point that it became clear that further assistance was required to make real progress. The Social Science Research Council kindly made a grant for this purpose. Meanwhile, Pat O'Sullivan, having done his Ph.D. on modelling freight flows in the Republic of Ireland, remained interested in this area of study and wished to take matters further. By one of those curious coincidences, he heard that the Ministry data were available to Michael Chisholm at about the same time that the latter learned that Pat O'Sullivan might be interested in joining the project. Our initial letters of enquiry crossed in the post.

The outcome was that Pat O'Sullivan came to Bristol, initially as a full-time Research Associate and subsequently as a lecturer on the teaching staff. The vacancy caused by Pat O'Sullivan's appointment to the permanent staff was briefly filled by Michael Worboys and then by Bob Dennis, to both of whom we are much indebted. We are also much in debt to Mr Derek Wagon in the Mathematical Advisory Unit of the Ministry of Transport for his assistance in providing the national freight data used in the study. The diagrams were all drawn by Mr. Simon Godden, Department of Geography, University of Bristol.

The work reported on in this book was made possible by a grant from the Social Science Research Council and by the Ministry of Transport making the basic data available. We wish to record our thanks to both organisations.

Michael Chisholm   Pat O'Sullivan
Bristol University   *Northwestern University*

NOTES OF GUIDANCE

Throughout the text, Imperial units have been used. In particular, note that all tonnage units are long tons.

In 1970, the Ministry of Transport was amalgamated with other government departments to form the Department of the Environment. To avoid confusion should the reader wish to check literature cited, we have used the title by which the department was known at the relevant date, i.e. Ministry of Transport or Department of the Environment.

To interpret coefficients of determination presented in the text, the reader may wish to note that with simple regressions run with $N = 78$, the 76×1 degrees of freedom mean that a value of F equal to or greater than 7.0 is acceptable at the ninety-nine per cent level. Thus values of R^2 of 0.09 or greater can be accepted as significant. We have, in fact, used the ninety-nine per cent level throughout this study.

Throughout the text, several symbols have been used frequently. To aid the reader, the more important ones are listed below, with a note on their meaning:

O_i	Total freight originating in zone i
D_j	Total freight terminating in (attracted to) zone j
d_{ij}	Distance between zones i and j
T_{ij}	Volume of freight moving between zones i and j
C_{ij}	Cost of transport between zones i and j
E_i and E_j	Activity levels in zones i and j
P_i and P_j	Population in zones i and j
a	Constant in regression equations
b	Regression coefficients
β	Gravity model distance exponent.

Chapter 1

INTRODUCTION

Transport accounts for somewhat over six per cent of the nation's Gross Domestic Product. The movement of freight is only a part of this total and it may therefore appear that aggregate freight costs are not a major element in the economy. Certainly, the evidence reviewed by Brown (1969) regarding the incidence of transport costs for industrial firms indicates that variations in these costs on account of location within the country are of little consequence. On the other hand, Edwards (1970a) estimates that for manufacturing industry as a whole the cost of transport accounts for at least nine per cent of the total cost of producing and distributing goods. With an average proportion approaching ten per cent of all costs, the implication is that spatial variations in transport costs are significant, at least for the manufacturing sector.

Given the apparent conflict of testimony there is need for further investigation. Some such work has already been reported by Chisholm (1971a); the present book takes the matter further and employs more elaborate analytical techniques. The importance of coming to a conclusion on this issue is that if spatial variations in transport costs are indeed important for firms and for regional economies, then this is a factor that must be taken into account in all government policy decisions regarding the location of employment and population. Conversely, if it is not important then attention should be directed to other considerations.

The second and related theme to which we have addressed ourselves is the fact that a large and continuing investment in transport facilities is necessary and that all such investment must necessarily be located in particular places. To have an adequate basis on which to make these decisions, it is necessary to have efficient methods for forecasting the volume of traffic that will move between nodes on the transport systems. This in turn depends upon the development and calibration of suitable models to give a good description of existing traffic flows. Although a great deal has been done at the intra-urban level, in the context of urban transportation/land-use models, only a handful of studies has been published dealing with freight flows at the inter-regional level in Great Britain. Two of the earlier studies were for particular regions of the country, north Kent (Starkie, 1967) and the Bristol region (Britton, 1967). The Ministry of Transport used a gravity model formulation to assess the spatial distribution of imports and exports, as one consideration germane to the decision whether Bristol should be allowed to proceed with major dock investments at Portbury (Ministry of Transport, 1966a). A more general attempt at modelling freight flows has been instituted at the Department of the

1

Environment (1971), though the results are not widely available. Following his work on the Republic of Ireland (1968 and 1969), O'Sullivan (1971) has also published some results obtained by applying the gravity model to freight flows in Great Britain. Clearly, surprisingly little work has been published concerning the modelling of inter-regional freight flows in Britain and there is in fact a great deal still to be done.

The main data sets available to us are based on seventy-eight zones of origin and destination covering the whole of Great Britain. This is a degree of spatial disaggregation that is substantially greater than is provided by the standard regions, for which road traffic data are available for 1962 (Ministry of Transport, 1966b). With only thirteen commodity groups, however, we are not able to approximate 'industries' as conventionally defined, except in one or two cases. For this reason, the analysis is pitched at a highly aggregate level in terms of commodities, with the consequence that our study is primarily relevant to the transport sector and for national policy relating to regional development rather than for specific industries or individual firms. Though we have related our experimental work to the relevant bodies of theory (chapter 2), our intention has been to exploit available data to further the positive understanding of how the space-economy works.

Thus, the context in which the present work is to be read is that of decisions by government regarding the location of investment in transport facilities and for the housing and employment of our growing population. While it would be rash to claim anything but a modest contribution in this field, it is also relevant to note first the dearth of previous comparable work and secondly the fact that there is no one perfect method of appraisal for major investments. Cost-benefit analysis, using discounted cash flows, is the most rigorous method available but is subject to a number of very serious limitations, as noted by Prest and Turvey (1965). It would be laborious to rehearse these difficulties at length, but two points are worth emphasising. With discounted cash flows, a time horizon much beyond twenty years is largely irrelevant if any kind of realistic rate of interest is to be used; at eight per cent compound, a benefit (or cost) of £100 twenty years hence has a present value of only £21.50. Since the time scale for major infrastructure investment – in transport facilities or in the urban fabric – normally exceeds two decades by a comfortable margin, the technique has serious limita tions within the data domain to which it applies. The second problem to note is that the data domain is only part of all relevant considerations which can be quantified, however notionally. The fuss attending the Roskill Commission's investigation of sites for a Third London Airport should be sufficient warning that agreement cannot always be obtained on the values to place on 'quantifiable' data, let alone the significance of unquantifiable items.

It follows that in practice major investment decisions must involve a good deal of judgement regarding disparate kinds of evidence. Consequently, there is need for a variety of approaches, a need underlined by the Severnside Study. This is the latest and most sophisticated of the major studies undertaken to examine the feasibility of substantial population growth in particular areas of the country. It is extremely lame on the question of both the capital and

2

recurring costs of the developments envisaged. Furthermore, no attempt is made to compare the costs associated with development on Severnside and the costs of a comparable scale of expansion elsewhere. With respect particularly to transport, the report is content to observe: 'Transportation investigations have been carried out only to the point where we could be reasonably satisfied that the proposed development possibilities are practicable' (Central Unit for Environmental Planning, 1971, paragraph 10.30).

The need for a variety of approaches to the questions raised in the preceding paragraphs is underlined by the great difficulties there are in interpreting the evidence of the 1963 census of manufacturing industry, the latest for which full details have been published. Examination of these data is deferred to chapter 9 and all we need note at this juncture is that at the scale of the standard region transport costs appear to be affected more by the size of regional population than by location within the country. Furthermore, the fact that 1963 is the last date for which full census of industry data are available lends some respectability to the use we have made of 1964 figures as the basis for our own study. However, there are some additional reasons why we have used material that is so dated.

As detailed in chapter 4, the first national freight survey was conducted in 1962. The results therefrom have been updated to 1964 and a further survey was put in hand for 1967—8. Results from this survey had not become available, even in November 1971, with the result that the 1964 data were the most recent at the time of our investigations. How serious a matter is it to base a study on a data source that will be approximately eight years out of date by the time this book is published? In part, the answer depends on the degree of stability possessed by the space-economy. In general, it seems abundantly clear that the spatial distribution of activities and of interaction between locations is remarkably stable over periods at least as long as a decade. Perhaps the most conclusive evidence for this proposition is the difficulty successive governments have experienced in making a real impact on regional imbalance through investment incentives, control of building and other means. Thus, it is reasonable to suppose that patterns of spatial organisation extant in the mid-1960s will be substantially unchanged even a decade later, despite the programme of roadway construction, railway modernisation, etc.

A more significant reason, however, why analysis based on data for 1964 has a general interest derives from the nature of the investigation. An important part of our aim has been to probe facets of the space-economy that have not been examined hitherto in this way. To some extent, therefore, the work reported in this book must be regarded as exploratory investigations which may serve two purposes: to point to the potential for work along similar lines given improved data; to invite comparison between findings for the British economy and for other nations that differ in geographical scale, spatial arrangement of resources and level of development (see Mennes, Tinbergen and Waardenburg, 1969, Appendix IV).

Except in chapter 9, attention is devoted exclusively to two modes of transport — road and rail. Inland waterways, coastal shipping and pipelines have

3

been ignored. While it is appreciated that this omission may affect the reliability of our findings, we plead in extenuation that it is only for road and rail that data on a comparable set of origins and destinations are available. Also, as table 1.1 shows, the two modes do in fact dominate the pattern of inland freight movements.

Table 1.1. *Great Britain: the distribution of freight traffic by mode of transport*

	1959	1964	1969
Million tons			
Road	1,146	1,400	1,570
Rail	234	240	206
Coastal shipping	43	52	48
Inland waterways	9	9	7
Pipelines	3	18	43
Total	1,435	1,719	1,874
Thousand million ton-miles			
Road	28.1	39.0	44.5
Rail	17.7	16.1	15.1
Coastal shipping	9.5	10.7	14.8
Inland waterways	0.2	0.1	0.1
Pipelines	0.1	0.7	1.6
Total	55.6	66.6	76.1

Source: *Annual Abstract of Statistics*, 1970, p. 217.

As implied in the previous paragraph, the unit of analysis is tons of freight. No consideration is given to the number of vehicle trips or wagon journeys required to move the freight flows. One reason for limiting the study in this way was the sheer volume of material to be handled and the consequent amount of work involved. The second and more respectable reason is that in making forecasts for investment purposes it is desirable to work from the most aggregated level towards the maximum degree of disaggregation deemed feasible. Such disaggregation may be based on various assumptions about modal split, size of unit loads, etc. Thus, the bench-mark forecast must be in terms of freight volume and all other sub-forecasts must be consistent therewith. This seems a sufficient reason for concentrating upon freight volume and ignoring the number of vehicle trips.

The book proceeds first to an examination of some basic concepts from location theory and interaction studies (chapter 2), to establish the framework of analysis. Theoretical nicety must be tempered by the practical necessities of data, as described in chapter 4, but a prior question is whether the space-economy of Great Britain can be regarded as sufficiently closed for the impact of foreign trade to be ignored, a problem discussed in chapter 3. Substantive results based on the Ministry of Transport data are presented in chapters 5, 6 and 7, covering respectively questions of traffic generation, traffic distribution and modal split.

Chapters 8 and 9 are relatively short and are devoted to an examination of two other data sets, in an attempt to measure transport costs. Finally, the substantive results are reviewed in the light of the theoretical framework developed in chapter 2.

In sum, the work reported in this book must be regarded as exploratory in nature, aimed at the improvement of inputs into cost-benefit analysis of investment proposals and also at the improvement of decisions based on the weighing of disparate evidence that cannot finally be entered into an economic balance-sheet. That the task is an important one is attested both by the lack of previous work on interregional transport flows and by the scale of investment decisions currently made. In the transport sector alone, infrastructure investment in roads, railways and sea ports amounted to £430 million in 1969 (Department of Economic Affairs, 1969). To this figure must be added the cost of regional policies, investment in New Towns, etc., making a total infrastructure commitment several times that for the transport sector alone.

Chapter 2

FRAMEWORK OF ANALYSIS

There are two main intellectual traditions that can be brought to bear on the empirical study of the spatial distribution of freight traffic and the associated question of regional comparative advantage. The choice of these two sources of ideas is conditioned partly by the nature of the available data, as described in chapter 4, and therefore the analytical modes adopted must not be regarded as the only ones that are either conceivable or feasible. On the one hand, partial equilibrium location theory is a fairly well articulated body of concepts that can be made to yield some testable hypotheses. On the other, general equilibrium theorising, while generally conducted at a rather rarified level of abstraction, includes a number of feasible, or potentially feasible, approaches to modelling the movement of goods. It is the purpose of this chapter to explore these two groups of ideas, and it will be convenient to start with partial equilibrium ideas from microeconomic location theory and thence proceed to more general models at the macroeconomic level. For the purpose of this discussion, the basic assumption is made that the economy of Great Britain can be treated as closed, i.e., that there are no trading relationships with oversea countries. This assumption is examined in detail in chapter 3.

Location theory

The habitual starting point for industrial location theory is the problem confronting a firm when making a decision as to the location of new investment. In the simplest version, this is conceived as the location of an entire plant (Weber, 1929); the problem is formulated as an optimisation problem, with an objective function to be either maximised or minimised. Weber assumed that the choice of location would not affect either the volume of sales or the unit revenue and consequently initially postulated that the optimum location is that which minimises the cost of transport, summed for both inputs and outputs. Lösch (1954) and Isard (1956) adopted the more defensible proposition that location does affect both sales volume and unit revenue and that, if firms are viewed as behaving in a rational economic manner, optimum location is defined as the point that yields the maximum profits. A further complication is introduced by the fact that a locational choice for the plant is interdependent with decisions about the location of sales, whether these decisions are taken by the firm itself or by its customers and potential customers. A third variable is introduced by the fact that firms may engage in a variety of price strategies, either as

6

alternatives to location strategies or simultaneously therewith (Isard *et al.* 1969; Chisholm, 1971b).

Firms are therefore viewed as seeking to achieve maximum profits. Hence, on the production side of the business, they are anxious to minimise the costs of operation, taking processing costs and the costs of assembling materials and distributing products. This proposition leads, if other things are equal, to the view that firms seek locations that minimise the total cost of transport incurred in their operations. Since it is generally assumed that transport cost is a monotonic function of distance (but see chapter 9), minimising transport costs is equivalent to minimising the total distance over which goods are moved. In other words, firms will seek to be as near to their suppliers and markets as they can: conversely, of the potential suppliers and customers, they will choose those that are nearest.

To the extent that there are several links in the productive chain, all the participants will have an interest in being located near each other. This is therefore one reason why firms agglomerate into major urban concentrations. Another reason, perhaps more important, lies in the generation of economies that are external to the individual firms. External economies of scale arise in a variety of ways, ranging from the wide choice of potential employees to the provision of common facilities such as transport and sewerage. Weber (1929) called these economies those of a 'labour orientation' and later economists (Myrdal, 1957) have elaborated the concept in the context of theories of economic growth.

Industries may be classified according as they are orientated towards their materials or their markets. At either end of the scale the distinction is clear, but there is in fact a sizeable proportion of industries for which the term 'mobile' (or footloose) would be more appropriate. More important in the present context, however, is the possibility that the sources of materials for manufacture and the markets for the products of a firm may be more or less co-terminous in space. In such a situation, the labels 'market-oriented' and 'materials-oriented' become meaningless. Yet, for a wide range of engineering industries, the spatial coincidence of material sources and market outlets is in fact a reality. Thus, one may suppose that firms will seek a location near to or at the centre of the space-economy. In this way, there could emerge a strong tendency for productive activity to become highly concentrated in a limited 'central' area, leaving much of the 'periphery' relatively poor and undynamic.

Potential accessibility

In an extreme form, the hypothesis may be stated as follows. Because firms seek to minimise the volume of transport inputs, they will tend to choose a location central in the space-economy. In operational terms, this means choosing a location with maximal access to the whole supply/market area. To define the accessible 'central' area, several writers have turned to notions of economic potential as a useful surrogate measure (Harris, 1954; Stewart & Warntz, 1958; Olsson, 1965). Derived from Newtonian physics, the concepts of energy and potential are not obviously and immediately transferable to economic and social phenomena. For the present purpose, what is required is a measure of the potential accessibility of any one place to all other places within the system. This may be regarded as a proxy for the facility with which the national market may be reached from each location. In Olsson's initial notation, the total potential of any location is given by:

$$i^v = G \sum_{j=1}^{n} \frac{P_j}{d_{ij}},$$

where i^v = the potential at place i

G = a constant

P_j = the population (or any other measure of 'mass') of city j

d_{ij} = the distance between i and j.

The expression above must be modified in two respects. First, it should include a term for the accessibility of place i to itself and therefore must include the case in which j is equal to i. Second, distance is given an exponent of one, which may be literally correct in terms of the analogy in physics but does not accord with the observed fact that there is a spatial elasticity of demand. Therefore, to measure the potential accessibility of locations, it may be appropriate to treat distance as raised to some power that can be assigned from empirically observed flow patterns. As is shown in chapter 6, the appropriate empirical distance exponent for road freight in Britain is -2.5 (i.e. $\frac{1}{d_{ij}^{2.5}}$), though most writers have chosen to use an exponent of -1.0. In applying the technique to Britain, Clark used estimates of transport cost instead of route miles but apparently otherwise adherred to the economic potential model in which the exponent is -1.0 (Clark, 1966; Clark, Wilson & Bradley, 1969).

There are other methods that may be used for measuring the centrality of a place (Ingram, 1971) including those derived from graph theory ideas of connectivity (Haggett & Chorley, 1969). However, the use of graph techniques results in the loss of a considerable amount of information and yields indices that are removed even further from the real world than are notions of potential accessibility. For this reason, no use has been made of graph theory techniques in the present study.

The concept of central versus peripheral locations and the relationship of this locational variable to profitability and economic growth has attracted a good deal of attention in the literature pertaining to both this country and others (for example, Peaker, 1971). It has therefore seemed worth exploring whether, at the level of aggregation determined by the traffic data available, the concept of potential accessibility has utility in explaining freight flow patterns. As indicated above, the choice of distance exponent poses a problem. Population-miles were obtained for all seventy-eight zones on the basis of both -1.0 and -2.5 distance exponents (see Appendix). The correlation between the two is high and positive ($R^2 = 0.74$), but not sufficiently high for us to feel justified in assuming that it is immaterial which is used. As the Appendix shows, the exponent of -2.5 yields a far larger coefficient of variation than does the

8

exponent of -1.0, indicating that the values are much more highly peaked in the former case than the latter. This suggests that a logarithmic transformation of $d_{ij}^{-2.5}$ values, the $d_{ij}^{-1.0}$ values being untransformed, might yield a higher correlation. In the event, the level of association is markedly reduced ($R^2 = 0.48$). The reason probably lies at least partly in the nature of the data. With an exponent of -2.5, the results are very sensitive to variations in the intra-zonal length of haul. As discussed on p. 34, the intra-zonal mean haul was obtained by a very crude technique. Close examination of the population-miles based on $d_{ij}^{-2.5}$ shows that the areas with very high values are all of small geographical extent, highly urbanised and either within or near major urban centres. Conversely, areas of large geographical extent all have medium or low values. This casts doubt on the utility of the exponent -2.5 in the present content and so reliance has been placed on population-miles using $d_{ij}^{-1.0}$.

One further preliminary exercise was undertaken to compare population-miles and employment-miles. Though the employed population is closely associated with the resident population in its spatial distribution, there are some differences in regional activity rates (Bowers, 1970), and in the vicinity of the conurbations commuting patterns upset the relationship between residence and workplace. With a distance exponent of -1.0, the value of R^2 between population-miles and employment-miles is 0.75. The greater part of the 'unexplained' variation can be attributed to commuting, especially around London, and is there-fore primarily a matter of compensating differences between adjacent zones. In principle, therefore, it is likely to be a matter of small moment whether population- or employment-miles are used as the measure of potential accessibi-lity. In the event, population-miles perform somewhat better than employment-miles as a surrogate for general accessibility (p. 72) and have therefore generally been preferred as the index of centrality.

Some testable hypotheses

If there really are significant differences in locational advantages for firms in different parts of the country, then the sum of location choices should produce a distinct spatial adjustment of activities. Without aspiring to an exhaustive list, the main kinds of spatial adjustment that we might hope to detect with suitable cross-section data for one year are:

(1) the profitability of enterprises: lower profits in peripheral areas;
(2) scale of enterprises: smaller ones in the peripheral areas;
(3) the mix of activities: proportionally smaller transport inputs in remoter areas;
(4) volume of freight: lower tonnages in the remoter areas;
(5) length of haul: to the extent that any or all of the above adjustments are inadequate, peripheral areas will experience longer mean hauls for freight generated and attracted.

Of the above five modes of adjustment, it is only the last two that can be directly examined with the transport data available to us. Such evidence as is

9

available on profitability yields an inconclusive answer: if there are regional differences in the profit levels achieved by firms, they appear to be too small to be detected (Brown, 1969; McCrone, 1969). As for the scale of enterprise, this is an elusive concept. There is in fact a marked concentration of small plants in the London and Birmingham regions, but these are often highly specialised in their production and serve a national market. There are no national data available to measure the geographical extent of market areas for plants located in various parts of the country. Nor are there data available bearing directly on the proportionate incidence of transport costs in various locations, apart from the sample studies reviewed by Brown (1969) and the 1963 census of manufacturing industries discussed in chapter 9. All of which leads us to suppose that items 4 and 5 above are likely to exhibit some spatial regularity related to the potential accessibility of the traffic zones.

In a nutshell, therefore, two hypotheses may be postulated:

(1) the greater the potential accessibility of an area, the larger the volume of freight per resident or per worker;
(2) the greater the potential accessibility of an area, the shorter the mean haul on freight traffic.

The two tendencies may, of course, operate simultaneously and may or may not be of sufficient magnitude to compensate the one for the other.

The second hypothesis is framed in terms of mean haul for freight movements. Other versions of this formulation are possible, notably that the β coefficients derived from a gravity model may be used instead of mean haul. The distance exponent or β coefficient is a measure of the rate at which interaction would decline with distance, given that other conditions are equal. On this reasoning, remote zones should try to avoid the costs of long hauls and will thereby have large distance exponents. A third formulation of the hypothesis has been experimented with, but proved unfruitful because of data problems. Imagine for any one zone the pattern of traffic destined to other zones; it may be conceived as an inverted cone. At a given distance, a certain proportion of the traffic will be accounted for; if we select a given proportion, say seventy-five per cent, then the cut-off distance for remote zones is likely to be greater than for accessible ones. However, initial experiments showed that this approach is not useful on the basis of the statistics available to us, for two related reasons. At the national level, about half the freight tonnage moves intra-zonally and for some zones the proportion rises to over ninety per cent. Thus, the cut-off distance is very sensitive to the proportion of intra-zonal freight and also to the estimated haul for this within-zone traffic. The cut-off distance is also sensitive to the substantial distance discontinuities that exist as the traffic from more distant zones is incrementally added to reach the critical value. Consequently, after some preliminary experiments, the cut-off concept was abandoned.

In the case of both of the above hypotheses, it would be desirable to disaggregate the total freight volume to examine for each zone the commodity-by-commodity variation. Unfortunately, too large a proportion of the cells in the commodity origin-destination matrices are empty, thereby introducing too many errors and

10

uncertainties for this line of enquiry to be pressed. However, it is possible to check whether the commodity mix of traffic for each zone has an appreciable affect on the mean haul.

An alternative approach to the question of regional comparative advantage was considered but rejected as the basis for formal analysis. If each area is conceived as an isolated economy engaging in 'foreign' trade with the other zones of the country, we may postulate two ways in which the degree of local self-sufficiency may vary. The smaller the total population of the zone, the more open its economy is likely to be. This is to be expected on all theoretical grounds relating to comparative advantage and the benefit of specialised productive activity and would accord with the general run of empirical evidence at the international level. On the other hand, an area that is remote from the main centres of population may seek to avoid some of the costs of transport imposed by long hauls by achieving a higher degree of local self-sufficiency than zones of similar population that are more centrally situated. On this basis, one would postulate that the proportion of traffic that is inter-zonal will vary inversely with the zone's population and the potential accessibility measured in population-miles. The fundamental snag with this approach is the variable shape and size of the traffic zones. As there is no satisfactory way of standardising the proportion of intra-zonal traffic for the size of reporting unit, there is no meaningful way in which to proceed.

There is, however, one subsidiary point that is worth noting and to which we shall return in chapter 3. A remote zone may compensate for the disadvantages of its location within Britain by concentrating on foreign trade to an extent greater than the national average. In this way, its relative disadvantage in terms of transport costs will be reduced or even eliminated. Such evidence as we can muster suggests that this does not happen, at least in terms of the volume of freight.

General models of the economy

The most satisfying and theoretically the most attractive approach to the relationships between flows of goods and the location of production and consumption is without a doubt the general equilibrium formulation. This stresses the interrelatedness of activities and defines an equilibrium state as one where all demands and supplies are reconciled as a set of uniquely soluble simultaneous equations. The difficulties of incorporating a spatial dimension into this scheme, if we work with a continuous geographical space identified by coordinates, themselves variables of the system, have not yet been overcome. The best that can be managed so far is a reduction of continuous space to a set of discrete regions connected by a network specification of the means of transport between them. We thus conceive of geographical space as a graph which we may manipulate as a matrix.

Given this simplifying assumption, it is possible to use operational forms of general equilibrium systems to model the movement of freight. The overall purpose of such models may be quite simply stated. If efficient models can be

11

developed that relate freight flows to one or more other variables, and if the relationships are stable over time, then it will be possible to make forecasts of freight flows given either forecasts or assumptions regarding the other variables in the system. This will be helpful in assessing investment requirements and long-term operating costs for future possible spatial distributions of activities. Furthermore, the derivation of models will throw light on the factors that influence the spatial distribution of freight flows and thereby on the advantages and disadvantages of different locations.

Several approaches are possible to the equilibrium modelling of freight flows and we shall therefore briefly consider the main lines of attack. Data limitations in fact rule out some of these possibilities, but it is nevertheless useful to include them in this discussion so that the operational forms actually used may be seen in perspective.

Linear programming

Activity analysis, or mathematical programming, enables one to establish the configuration of a system which maximises or minimises some objective within a domain of feasible solutions. The most widely used method is that of linear programming, in which it is normally assumed that the relationships considered are linear; on this assumption, it is possible to use algorithms such as the SIMPLEX method to find the optimal solution. In the present context, it is the Transportation Problem version of linear programming that is relevant, in which the objective is to minimise the cost of transport, or some surrogate for cost such as ton-miles.

As an apt illustration, consider a simple one-commodity, many-region economy, in which the objective is the minimisation of the cost of production and transportation. If we know:

C_{ij} = cost of production in region i plus transport to j,

O_i = quantity of the good produced in i,

D_j = quantity of the good required in j,

then the objective is to find a set of flows between i and j such that:

$$\text{Min } C = \sum_i \sum_j T_{ij} \cdot C_{ij}$$

subject to $\sum_j T_{ij} \leqslant O_i$, production capacity not being exceeded in any i

$\sum_i T_{ij} \geqslant D_j$, all requirements being satisfied

assuming total capacity exceeds total demand $(\sum_i O_i > \sum_j D_j)$ and

$$T_{ij} \geqslant 0$$

the solution values being non-negative. The maximising *dual* to this minimising *primal* problem is:

$$\text{Max } V = \sum_j V_j \cdot D_j - \sum_i U_i \cdot O_i,$$

where U_i and V_j are shadow prices associated with the origins and destinations respectively. The maximisation is subject to

$$U_i - V_j \leqslant C_{ij}$$

there being no excess profits (the perfect market equilibrium condition) and

$$V_j, U_i \geqslant 0,$$

the shadow prices of the dual solution being non-negative.

This simple Transportation Problem may be generalised to include many commodities and sectors, including transport, government and the rest of the world, and to allow for substitutability of inputs between various sectors. However, in this more generalised form acute problems arise over data on costs and prices, inter-industry relations and the degree of substitution that is feasible. Consequently, it is more usual to use the simpler versions, which make much less serious demands on the supply of information. In particular, if production capacity at each origin and the demand at each destination are given and are assumed to be perfectly inelastic, i.e., not to vary in response to changes in prices and costs, then the simplest version of the Transportation Problem is that which assigns the flows in such a way as to clear all supplies and satisfy all demands for the least cost of transport. If transport costs are a monotonic function of distance, then the problem reduces to minimising the ton-miles of traffic for the whole system.

The Transportation Problem is, of course, constantly being solved in order to make commercial decisions. Henderson (1958) applied the method to an analysis of coal movements in the United States and Land (1957) analysed coking coal movements in Britain in a similar manner. A more elaborate formulation was used by Heady and Skold (1966) to forecast an optimal pattern of location of agricultural production and trade for several crop combinations and products.

Input-output

Another mode of general equilibrium operation is the construction of an input-output system. If we have observations of the flows of transactions between sectors of the economy and divide these by the total outputs of the sectors for which they are destined, we obtain a matrix of 'technical input-output coefficients' (A). Subtracting this from an identity matrix (I) and inverting gives us a means for calculating all the direct and indirect interindustry effects of any change in demand for final goods. So, if Y is a vector of final demands and O of total industry outputs:

$$(I - A)^{-1} Y = O.$$

Given any set of future final demands and a matrix of input-output coefficients, and assuming them to be stable over time, we can forecast future outputs by sectors.

This structure can be further disaggregated by regions so that our transactions can be analysed as flows between regions as well as sectors of the economy. If

13

we had direct observations of these flows we could construct an interindustry, interregional input-output coefficient matrix (B) and manipulate it as before, specifying outputs by regions:

$$(I - B)^{-1} Y = O_i.$$

To collect data for interindustry, interregional flows by direct observation is extremely expensive. Even if information is obtained by means of sampling freight consignments, and regional disaggregation is confined to the ten planning regions covering Great Britain, it would cost between £0.25 and £0.5 million merely to acquire the data (Edwards & Gordon, 1971, 425). Consequently, some method of estimation must be resorted to which will involve dubious assumptions. For example, Chenery, Clark & Cao-Pinna (1953) assumed some constant proportionally between the amounts of a good imported from another region and amounts provided internally for production to calculate 'interregional trade coefficients'. The most ingenious method of dealing with this problem is that of Leontief and Strout (1966), which overcomes the rigidity of the trade coefficients by assuming that users and suppliers are indifferent to sources and destinations, responding only to prices. They suggest that interregional trade in any one commodity can be estimated by a set of structural equations as follows:

$$T_{ij} = \frac{O_i \cdot D_j}{T} \cdot Q_{ij} \text{ for } i \neq j$$

where T_{ij} = the flow from origin to destination j,

T = the total flow in the economy, $\sum_i \sum_j T_{ij}$,

O_i = the total amount flowing out of i, $\sum_j T_{ij}$,

D_j = the total amount flowing into j, $\sum_i T_{ij}$,

and Q_{ij} = empirical constant reflecting the cost of movement between i and j and their relative competitive positions.

However the input-output problem is approached, it is evident that an interregional system, even with a limited number of regions, makes enormous demands upon data, demands that far outrun the present or the prospective supply. As a consequence, the number of British regional input-output studies actually published is small and the majority are open to severe criticism, especially where the number of regions exceeds three (Edwards & Gordon, 1971). In the present context, therefore, the input-output approach must be rejected as not feasible.

Gravity model

Where we have limited knowledge of the way the system works and also possess only crude data — the two propositions are of course closely connected — we may resort to the methods of social physics to develop a model of traffic flows.

If we know only

T = total tonnage of any commodity,
O_i = tons generated by i,
D_j = tons attracted by j,
and d_{ij} = distance between i and j

and, for the sake of simplicity, assume $T = \sum_i O_i = \sum_j D_j$, Wilson (1967) has shown how we can derive an expression for a 'most probable distribution' of flows (T_{ij}). We can write down constraints on these T_{ij} delimiting a feasible domain, representing ways in which we know the behaviour of the system to be non-random:

$$\sum_j T_{ij} = O_i, \qquad (1)$$

$$\sum_i T_{ij} = D_j, \qquad (2)$$

and, assuming that cost of movement varies as the log of distance:

$$\sum_i \sum_j T_{ij} \log d_{ij} = C. \qquad (3)$$

(1) and (2) are the constraints on availabilities and (3) fixes total expenditure at any time on moving this commodity to a constant sum, C. The number of distinct individual arrangements, $w\,(T_{ij})$, of ton flows which give rise to a distribution T_{ij} is given by:

$$w\,(T_{ij}) = \frac{T!}{\prod_{i\,j} T_{ij}!}$$

The total number of possible distinct arrangements of individual tons is then: $W = \sum w(T_{ij})$, where the summation is over all distributions which respect (1), (2) and (3). It can be assumed that the maximum $w(T_{ij})$ dominates the other terms of this sum to such an extent that the distribution which gives rise to it is overwhelmingly the most probable. So we have to find an expression for T_{ij} which will maximise $w(T_{ij})$ subject to (1), (2) and (3). To do so the following is maximised:

$$M = \log w + \sum_i \lambda_i(O_i - \sum_j T_{ij}) + \sum_j \lambda_j(D_j - \sum_i T_{ij}) + \beta(C - \sum_i \sum_j T_{ij} \log d_{ij}),$$

where λ_i, λ_j and β are Lagrangean multipliers. It is convenient to maximise $\log w$, as it is then possible to use Stirling's approximation to estimate factorial terms, and any monotonic function of w would give the same answer. The most probable distribution of tons is given by the solution of:

$$\frac{\partial M}{\partial T_{ij}} = 0.$$

Stirling's approximation gives:

$$\frac{\partial \log N!}{\partial N} = \log N$$

15

and so

$$\frac{\partial M}{\partial T_{ij}} = -\log T_{ij} - \lambda_i - \lambda_j - \beta \log d_{ij}$$

and this becomes zero when

$$T_{ij} = \exp(-\lambda_i - \lambda_j - \beta \log d_{ij}).$$

To get λ_i and λ_j we substitute into (1) and (2)

$$\exp(-\lambda_i) = \frac{O_i}{\sum\limits_i \exp(-\lambda_j - \beta \log d_{ij})} \quad \text{and} \quad \exp(-\lambda_j) = \frac{D_j}{\sum\limits_j \exp(-\lambda_i - \beta \log d_{ij})}$$

To obtain a more familiar form, write:

$$\frac{\exp(-\lambda_i)}{O_i} = A_i \quad \text{and} \quad \frac{\exp(-\lambda_j)}{D_j} = B_j,$$

then

$$T_{ij} = A_i . B_j . O_i . D_j . d_{ij}^{-\beta},$$

where

$$A_i = (\sum\limits_j B_j . D_j . d_{ij}^{-\beta})^{-1}$$

and

$$B_j = (\sum\limits_i A_i . O_i . d_{ij}^{-\beta})^{-1}.$$

This is known as the production and attraction constrained gravity model. Wilson (1968) has shown how this method can be extended to incorporate the Leontief-Strout equations as constraints to produce an integrated model.

A crude version of the gravity model has been developed to examine the effects of changes in the transport system — notably, the provision of new routes or improvements on existing ones. The 'abstract mode' approach of Quandt & Baumol (1966) as developed by *Mathematica* (1968) in the North East Corridor Study is designed for just this purpose. The demand for freight between two regions by transport mode m, T_{mij}, is assumed to depend on:

(1) the population of the regions, P_i and P_j,
(2) the gross regional products, Y_i and Y_j,
(3) indices of the industrial character of the two, M_i and M_i,
(4) the least shipping time H_{ij}^b and relative travel time for the mth transport mode, H_{mij}^r,
(5) the least cost of shipping C_{ij}^b and the relative cost for the mth mode for a certain commodity C_{mij}^r,
(6) the number of modes serving i and j, N_{ij}.

These relationships were formulated in such a way as to yield an estimating

16

equation which is conveniently linear in the logarithms of the variables, thus:

$$T_{mij} = \alpha_0 . P_i^{\alpha_1} . P_j^{\alpha_2} . Y_i^{\alpha_3} . Y_j^{\alpha_4} . M_i^{\alpha_5} . M_j^{\alpha_6} . N_{ij}^{\alpha_7} . f_1(H) . f_2(C),$$

where

$$f_1(H) = (H_{ij}^b)^{\beta_0} (H_{mij}^r)^{\beta_1}$$

and

$$f_2(C) = (C_{ij}^b)^{\gamma_0} (C_{mij}^r)^{\gamma_1}$$

The advantage of this model is that it allows a forecast to be made of the effect of a change in the transport system or a new mode simply by specifying new costs and shipping times – hence the appellation 'abstract mode'.

The data available to us permit the use of the gravity model in the form developed by Wilson, but do not permit us to go on to the 'abstract mode' version.

Behaviour of the firm

As discussed on pp. 6–11, location theory is a partial equilibrium analysis in microeconomics. If we return to this arena, an interesting extension of the argument can be developed. We assume that firms behave in a normative profit-maximising manner. We may then adopt the reasoning employed by the *Mathematica* team (1968) and start with a cost function for transport

$$C = rT + stT + k\sqrt{(s + t)} . T \qquad (1)$$

where C = total cost per annum of handling the product,

     T = total tonnage shipped in a year,

     r = shipping cost per unit between a given origin and destination,

     s = time between shipments,

     t = average time required to complete a shipment,

and k = some constant.

So we have the unit rate times amount shipped, plus in-transit costs plus an element representing the level of stock which the shipper has to maintain as an insurance against mishaps – his inventory. To obtain an expression for marginal cost as a function of total annual tonnage, we take the partial derivative of (1) with respect to T:

$$\frac{\partial C}{\partial T} = r + st + k\sqrt{(s + t)}. \qquad (2)$$

The reason goods are shipped between regions is that a price difference exists. The supplier's total revenue will be the amount shipped times the price difference (Δp) between the two locations, $\Delta p.T$. His marginal revenue will be given by:

$$\text{M.R.} = \Delta p + T \frac{\partial \Delta p}{\partial T}, \qquad (3)$$

17

where the second term indicates the elasticity of demand at the destination. It can be assumed that the demand function is linear:

$\Delta p = a - bT$, in which case (3) becomes:

$M.R. = \Delta p - bT$.

To obtain the equilibrium position marginal cost must be equated to marginal revenue:

$\Delta p - bT = r + st + k\sqrt{(s + t)}$.

This can be rearranged to give an equation explicit in T:

$$T = \frac{1}{b}[\Delta p - r - st - k\sqrt{(s + t)}]. \qquad (4)$$

This says, quite sensibly, that the annual tonnage shipped is larger the larger the price difference, the smaller the freight rate, the smaller the time between shipments and the smaller the slope of the demand curve.

This equation can be translated into an estimating equation by the addition of a stochastic component and the identification of origins and destinations (i and j), commodities (k), modes (m) and years (t) by subscripting variables:

$$T_{ijkmt} = \alpha_0 + \alpha_1 \Delta p_{ijkt} + \alpha_2 r_{ijkmt} + \alpha_3 (s_{jkt} \cdot t_{ijkmt}) + \alpha_4 (s_{jkt} + t_{ijkmt})^{\frac{1}{2}}.$$

The reader will have noted that though the formal structure of this model is quite simple, it nevertheless again makes heavy demands on the data supply — cost of transport, time between shipments and time in transit, etc. Unfortunately, for the national system of freight movements these data demands cannot be met.

Summary

Of the general equilibrium models reviewed in this section, the recurring problem is the extent to which the data needs outrun the existing, or even the potential, supply. Consequently, we are driven back to the simpler models, in particular to simple transportation problem linear programming formulations and the gravity model. Though these depend on the veracity of a very limiting set of assumptions, they are, nevertheless, the best approaches for modelling freight flows given the kind of data actually available. In practice, they do permit us to develop models that are reasonable approximations to reality and which enable us to probe the structure of the British space-economy in a way that has not hitherto been done.

Conclusion

From this review of possible approaches to the freight aspects of the British space-economy, some relatively simple hypotheses regarding location have emerged as amenable to examination and so have two major ways of modelling the national freight flows. The hypotheses are that the more peripheral an area,

18

the smaller the volume of freight *per caput* and/or the longer the mean haul. The alternatives of the 'most probable' distribution of the gravity model and the 'optimal' solution of linear programming have been applied to the modelling of traffic flows. Both of these are partial equilibrium solutions to the problem, in which it is assumed that the distributions of supply and demand are given, as well as the cost of transport. The two models bear some structural resemblance to each other. However, they differ in the assumptions they make regarding the degree of perfection in the behaviour of the actors in the system. The programming model assumes precise, optimising behaviour with perfect knowledge, while the gravity model uses an empirically estimated function to describe behaviour with respect to cost and distance. The general form of the distance attenuation function derives from a seemingly harmless constraint on the behaviour of the system — that the total amount spent on transport in any one time period is a fixed amount. Both approaches were tried in the present context, because it was hard to be sure to what extent the assumption of optimising behaviour embodied in the linear programming approach was tenable, or, if tenable, observable, given the heterogeneous nature of the commodity classes for which data were available.

The fundamental strategy adopted in the present study is similar to that used in urban land-use/transportation studies. In this staged approach, the first task is to establish some basis for forecasting the volume of freight generated by and attracted to each zone and the second is to develop efficient models for predicting the spatial distribution of the resulting flows. To forecast the generation and attraction of freight, experiments have been performed with population, employment and retail expenditure as independent variables. For both the generation and the distribution of traffic, the procedure has been to start with the most highly aggregated freight flows and then to disaggregate by mode of transport and commodity group. However, before we can proceed to this question, it is necessary to revert to a point made at the beginning of this chapter: is it justifiable to regard the economy as closed for the purpose of modelling freight flows and examining the associated costs? It is to this question that we turn in chapter 3.

Chapter 3

THE PROBLEM OF FOREIGN TRADE

The British economy is usually regarded as an 'open' one, with external transactions playing a major part in the nation's activities. Since visible trade, defined
as the sum of imports and exports by value, amounted to £15,661 million in
1969, whereas the Gross Domestic Product in that year was £38,150 million,
it is abundantly clear that our livelihood is intimately bound up with foreign
transactions. In terms of money flows, therefore, it would be patently absurd
to regard the economy as a closed one. Fortunately for the study of freight
movements within the country, the same conclusion does not apply in terms
of freight volume. There are several important reasons why this is so. The unit
value of imports and exports tends to be considerably higher than the unit
value of goods traded domestically. In addition, some foreign trade does not
enter the internal land transport system for which we have data. This is especially
the case with petroleum products moved by pipeline from coastal refineries and
electricity generated from coastal oil-fired power stations. The third reason is that
some imports, notably iron ore, are processed at or very near the point of import
and suffer a substantial reduction in bulk in the process. The present chapter is,
therefore, devoted to a discussion of the evidence relevant to the proposition
that for the present purpose the economy of Great Britain may be treated as
closed, isolated from the world at large.

Table 3.1 shows that in 1969 inland traffic amounted to 1,874 million tons,

Table. 3.1. *Great Britain: inland and oversea traffic movements, million tons*

	Inland					Oversea		
Year	Total	Road	Rail	Coastal shipping, mean of inwards and outwards movements		Total	Imports	Exports
				1	2			
1969	1,874	1,570	206	48	54	225	183	43
1968	1,851	1,550	207	48	53	214	173	41
1967	1,788	1,500	201	49	54	197	159	37
1966	1,757	1,450	214	54	56	192	155	38
1965	1,747	1,430	229	54	56	186	151	35

Sources: *Annual Abstract of Statistics* (various issues) and National Ports Council, *Digest of Port Statistics*.
[1] As published in the *Annual Abstract of Statistics*.
[2] As published in the *Digest of Port Statistics*.

20

The problem of foreign trade

whereas oversea trade summed to only 225 million tons in the same year. The greater part of overseas freight probably does enter inland movements, as goods distributed to and from the docks. Thus, as a first proposition we may say that approximately twelve per cent of the freight tonnage moving within Great Britain has an origin or destination abroad.

In recent years, over half the tonnage of our foreign trade has consisted of fuels, mainly imports of crude petroleum and exports of refined petroleum products; in the period 1965–9, the proportion was 55.9 per cent (table 3.2).

Table 3.2. *Great Britain: commodity composition of foreign trade, million tons*

	Imports					Exports			
Year	Total	Fuels	Food-stuffs	Basic materials	Manufactured goods	Total	Fuels	Food-stuffs	Basic materials	Manufactured goods
1969	183	112	20	37	14	43	19	2	5	16
1968	173	103	19	37	14	41	18	2	5	16
1967	159	94	19	33	13	37	15	3	5	14
1966	155	91	19	34	11	38	17	3	4	14
1965	151	82	19	38	11	35	16	2	4	14

Note: owing to rounding, the sum of the commodities does not always exactly agree with the total shown.
Source: National Ports Council, *Digest of Port Statistics* (various issues).

Unfortunately, data on the weight of imports and exports are not available for earlier years, with the exception of 1961 (Ministry of Transport, 1962), whereas the latest year for which we have a commodity breakdown of inland traffic is 1964. Thus, if we assume that in 1964 roughly half the nation's foreign trade was fuels, the non-fuel component amounted to about ninety million tons; this is approximately eight per cent of inland freight other than fuels in that year. On the other hand, foreign trade in fuels was equivalent to half the volume of inland traffic in coal and petroleum.

With the major exception of fuels, it appears reasonable to conclude that external trade is not large enough in aggregate volume seriously to disturb the analysis of traffic flows within Britain.

Port zones compared with other zones

However, the same is not necessarily true of all regions. There is some evidence, for example, that Scotland contributes proportionately more to exports than does the rest of the nation (McCrone, 1969). Only one study has been carried out in this country that links freight passing through individual ports and the inland origin. This was a sample study of the inland origin of exports in 1964, known as the Martech study. Though results have been published (National Ports Council, 1966), the data have been very severely criticised on several grounds and are thought to be too unreliable to be used. In the absence of

21

Freight flows and the British economy

reliable data on the inland origins and destinations of foreign trade, it is possible only to compare the volume of seaborne freight with the volume of inland traffic for each of the Ministry zones that has one or more ports. Owing to processing and manufacturing operations in or beside the dock, some of the imports will not appear directly as originating inland traffic; similarly, some exports may not represent terminating inland traffic. Therefore, the propositions which can be made for zones with ports are as follows:

(1) some of the freight landed at a port subsequently appears as originating (O_i) inland traffic,

(2) some of the freight exported from a port has previously appeared as terminating (D_j) inland traffic.

There is no means of knowing how much foreign trade does appear in the inland freight movements, nor of telling what proportion of that which does either originates or terminates in the port zone itself. However, if the inward seaborne freight is calculated as a percentage of the originating inland traffic for the port zone and a similar calculation is made for the importance of exports relative to terminating traffic, the orders of magnitude will indicate whether foreign trade is likely to be a significant part of the activities of the various coastal zones. The data are shown in table 3.3.

The allocation of ports to the Ministry of Transport zones is straightforward in most cases. However, ambiguity does occur in some cases, especially London. The Port of London Authority includes the docks at Tilbury, which are in zone 47, while the docks up-river are spread among several zones. All London's traffic has been assigned to zones 48, 49, 50 and 54 collectively. Similar but lesser problems occur in respect of Merseyside, Clydeside and Tyneside. Consequently, the attribution of the port traffic to the inland traffic zones is only approximate and in some cases may be subject to considerable error. A further but minor error is introduced by the fact that the earliest year for which the tonnage of freight passing through ports is available is 1965, whereas the inland freight data refer to 1964.

Table 3.3 displays a striking contrast between the relative importance of imports and exports. Imports totalled 151.0 million tons, whereas exports amounted to only 35.2 million tons; it is hardly surprising, therefore, that on average for the port zones shown, imports are four times as important as exports relative to inland freight. More interesting is the much greater variation in relative importance of imports than of exports — compare the range of 1.0–78.2 per cent for imports and 0.0–17.6 for exports. Part of the explanation for this contrast lies in the high proportion of imports accounted for by fuels, mainly petroleum (table 3.2), the greater part of which import trade is accounted for by eight ports. A comparison of tables 3.3 and 3.4 shows that fuel imports range from about one half to over ninety per cent of total imports into the relevant Ministry zones. This raises a serious problem of interpretation, since crude petroleum is generally refined at or near the point of import, and the refined products are then either shipped out by coastal vessels or dispatched inland. In the latter case, rail and road transport may be used and such move-

22

Table 3.3. *Comparison of inland (1964) and oversea (1965) freight movements for all port zones*

Ministry of Transport zone	Inland freight (million tons)		Oversea freight (million tons)		col.4 as % col.2	col.5 as % col.3
	O_i	D_j	Imports	Exports		
1	2	3	4	5	6	7
66	19.4	18.7	15.2	1.5	78.2	8.1
34	26.7	28.5	19.9	4.9	74.6	17.6
56	31.5	32.8	13.4	0.6	42.4	1.7
31	22.1	25.6	8.6	1.6	40.0	6.3
72	23.7	26.2	9.4	0.9	39.6	3.3
8, 9	20.7	30.3	7.4	1.6	35.6	5.2
48, 49, 50, 54	108.7	106.6	29.8	5.1	27.5	4.8
55	63.5	66.8	13.1	5.7	20.7	8.8
17	21.5	23.5	4.3	1.2	20.2	4.9
39	22.6	27.6	3.9	0.9	17.4	3.3
65	80.6	78.8	9.3	4.0	11.5	5.2
21, 22	13.4	16.6	1.4	0.2	10.3	1.4
18	10.0	9.5	1.0	0.1	10.2	0.6
2, 3	22.8	24.6	2.3	0.8	10.1	3.2
70	16.0	12.2	1.2	1.4	7.4	11.2
58	60.3	61.6	4.4	0.4	7.3	0.7
76	17.5	15.0	1.0	0.1	5.6	0.8
11	10.3	10.3	0.5	0.3	4.9	2.8
45	43.7	46.9	1.5	0.5	3.4	1.0
5	4.2	5.0	0.1	0.2	3.0	4.0
74	14.2	13.3	0.4	0.1	2.9	0.6
68	34.9	38.6	0.8	0.2	2.3	0.4
69	22.5	19.4	0.4	0.2	2.0	0.8
67	17.5	18.6	0.3	0.2	1.5	0.9
59	31.7	34.1	0.4	1.8	1.3	5.2
47	36.2	35.1	0.4	0.3	1.2	0.8
1	12.2	9.6	0.1	0.2	1.0	1.6
10	8.3	8.4	0.1	−	1.0	−
40	42.5	32.1	0.4	0.2	0.9	0.7
Total	859.2	876.3	151.0	35.2	17.6	4.0

Sources: Inland freight, see chapter 4. Oversea freight, National Ports Council, *Digest of Port Statistics,* 1966.
Note: the figures have been rounded to one decimal place after calculation. The traffic zones are identified in the Appendix.

ments will appear as inland originating (O_i) traffic. However, if pipelines are used for refined products, as from Fawley, these movements escape notice in the data sets available to us. Milford Haven presents a particularly good example of the problems involved. The Haven is a terminal for a crude oil pipeline to the Llandarcy refinery near Swansea. Some of the refined products from the

Freight flows and the British economy

Table 3.4. *Oversea trade in fuels, 1965, for individual major ports*

Port	Ministry of Transport zone	Imports (million tons)	Exports
London	48, 49, 50, 54	16.9	0.8
Milford Haven	66	15.1	1.5
Southampton	56	12.3	0.3
Medway	55	10.9	5.5
Liverpool	34	10.2	0.4
Clyde	72	4.6	–
Manchester	31	4.4	0.6
Bristol	58	1.0	0.1
Swansea	65	0.7	2.2
Total		76.1	11.4

Source: National Ports Council, *Digest of Port Statistics,* 1966.

Llandarcy refinery are dispatched by pipeline to Swansea docks for coastwise shipment, some are similarly sent to a nearby gas manufacturing plant and a plastics plant, while some are also dispatched by road. But Milford Haven is also the site of refining activities, with refined products sent out by road and rail as well as coastwise.

At least some of the foreign trade must represent goods drawn from a wide hinterland or destined thereto. For this reason, it would be expected that the volume of freight originating or terminating in the port zones would be disproportionately large and have movement characteristics different from the non-port zones. In terms of tonnage per resident, table 3.5 shows that in fact

Table 3.5. *Traffic characteristics of port zones and all other zones, median values*

		Port zones (N = 35)	Other zones (N = 43)
Tons of freight per resident			
Road plus rail	O_i	35	38
	D_j	35	33
Road only	O_i	29	31
	D_j	27	31
Mean haul, miles			
Road plus rail	O_i	37	31
	D_j	38	32
Road only	O_i	32	29
	D_j	33	28
Gravity model β coefficients, road only	O_i	−2.7	−2.3

there is no substantial excess compared with other zones: in fact, the reverse situation seems to obtain to a small extent. A formal test of differences between means is not appropriate, given the very disparate weighting to be attached to the zones. As the differences in the median values are well within the range of variation of tonnage per resident, it is plausible to regard the differences as not significant. The differences on mean haul are somewhat greater and are more consistent; port zones do have a somewhat longer average haul, which is consistent with the idea of servicing a wide hinterland. On the other hand, the port zones also show a greater distance friction, as measured by the gravity model β coefficients. This implies that, were other things equal, the port zones would be characterised by a greater distance friction than other zones, which is contrary to what one would expect. The interpretation to be placed on this apparent contradiction is that the port zones tend to be peripheral to the main centres of population: their median value for population-miles is 474,000, compared with 666,000 for the other 43 zones. Indeed, the relationship between mean haul and the β coefficients for the dichotomous grouping accords with the finding in chapter 6 that though peripheral zones try to avoid the costs of long hauls, as indicated by large β coefficients, they do in fact experience longer mean hauls, because the adjustment is incomplete. This strongly suggests that the inland traffic of port zones as a group does not have special characteristics differentiating it from that of other traffic zones.

Comparison of the port zones

One point is abundantly clear from table 3.3. It is only in the case of the zones containing major ports that foreign trade is substantial relative to inland traffic. Imports as a percentage of originating (O_i) traffic in general exceed ten per cent only for those zones with at least 4.0 million tons of imports. Similarly, an annual export of at least 1.0 million tons is necessary before exports approach five per cent of the terminating (D_j) inland freight. The clear presumption is that zones in peripheral localities with small local ports do not in fact engage in foreign trade to a degree above the national average, at least in terms of freight tonnage, unless their trade is in fact conducted through ports other than those in the zone itself. If zones such as 66 (West Wales) and 68 (North Scotland) do engage in foreign trade to an extent above the average, it seems that this must be through ports elsewhere or by virtue of concentrating on goods of high unit value. A net conclusion relevant to the present study is that the main impact of foreign trade on the volume of inland freight flows is confined to the zones with the major ports.

If the port zones as a class do not display traffic characteristics that differentiate them from the other zones, it may be that some systematic pattern can be distinguished within this group of 35 zones. Because foreign trade is such a variable proportion of inland freight movements (table 3.3), is there evidence to show that where foreign trade is relatively important the inland traffic characteristics differ from those zones where oversea trade is of little significance? The answer is negative. For the port zones listed in table 3.3, the proportion of

Freight flows and the British economy

oversea trade to inland traffic does not vary systematically with the volume of oversea trade. Nor do the summary measures used in table 3.5 bear any relationship to the size of a port zone's foreign traffic.

Thus within the group of port zones there is no detectable impact of foreign trade related to the volume of the oversea transactions. This conclusion reinforces the point that, as a class, the port zones are not distinguished from the other zones in terms of their traffic characteristics.

Foreign trade and the balance of inland traffic O_i and D_j

The rather surprising conclusion of the two previous sections is supported by other evidence. For the national economy as a whole, the volume of inland freight originating (O_i) must equal the volume of terminating freight (D_j). For each part of the system, however, no such equality need exist. Imports in 1965 totalled 151.0 million tons, whereas exports amounted to 35.2 million tons. An imbalance of this magnitude would be expected to affect the ratio of O_i/D_j traffic in the port zones; one would expect the aggregate O_i volume to exceed the D_j volume by anything up to 110–20 million tons. As table 3.3 shows, the O_i traffic of the port zones amounted to 859.2 million tons, whereas D_j freight added up to 876.3 million tons. The difference of 17.1 million tons is small and cannot be treated as very significant but it is in the *opposite* direction to that which would be expected. There is thus a discrepancy of between 127 and 137 million tons, representing an apparent deficit of O_i traffic relative to D_j freight in the port zones (17 million tons plus 110–20 million tons). If the oversea traffic in fuels is eliminated from consideration, 1965 imports, at 69 million tons, exceeded exports by 50 million tons. When inland movements of fuel are also excluded, the port zones' surplus of D_j traffic of 17 million tons is converted into a small deficit of about 18 million tons (table 3.7). Even with fuels excluded, there is therefore still a substantial discrepancy whereby the port zones apparently consume 32 million tons of freight (foreign trade and inland traffic) more than they generate.

As table 3.7 shows, the aggregate net surplus of terminating traffic in the port zones is composed of two contrary elements: road traffic shows a small deficit of 6.1 million tons, whereas terminating rail freight is in surplus to the extent of 23.1 million tons. This suggests something which has already been alluded to in previous paragraphs, namely that the commodity structure may be an important consideration. Table 3.7 shows that by far the most important commodity with an excess of D_j over O_i traffic is coal and coke, with most of the surplus carried by rail. The next most important item is building materials, in this case with road transport responsible for most of the excess. Very little of either commodity will be exported and the large quantity of these commodities required in the port zones therefore reflects the activities located there and the shortage of coal and building materials within the port zones. The deficit in D_j traffic occurs mainly in petroleum products and foodstuffs, and road transport is the means of carriage mainly responsible. For both commodities, but especially with petroleum, large quantities are imported and therefore

26

Table 3.6. *Traffic of port zones, million tons*

Ministry of Transport zone	Oversea freight 1965, million tons			1964 Inland freight, excess (+) or deficit (−) of D_j over O_i million tons		
	Imports	Exports	Imports minus Exports	Road	Rail	Total
48, 49, 50, 54	29.8	5.1	24.7	−5.0	+3.1	−1.9
34	19.9	4.9	15.0	−0.5	+2.4	+1.9
66	15.2	1.5	13.7	+0.9	−1.6	−0.7
56	13.4	0.6	12.8	+1.1	+0.2	+1.3
55	13.1	5.7	7.4	+2.9	+0.3	+3.2
72	9.4	0.9	8.5	+2.9	−0.4	+2.5
65	9.3	4.0	5.3	−1.4	−0.5	−1.9
31	8.6	1.6	7.0	+1.9	+1.6	+3.5
8, 9	7.4	1.6	5.8	−0.3	+9.9	+9.6
58	4.4	0.4	4.0	−0.1	+1.3	+1.2
17	4.3	1.2	3.1	−1.0	+3.0	+2.0
39	3.9	0.9	3.0	−3.4	+8.5	+5.1
2, 3	2.3	0.8	1.5	+1.5	+0.2	+1.7
45	1.5	0.5	1.0	+1.7	+1.4	+3.1
21, 22	1.4	0.2	1.2	+1.2	+2.1	+3.3
70	1.2	1.4	−0.2	−4.1	+0.2	−3.9
18	1.0	0.1	0.9	−0.2	−0.3	−0.5
76	1.0	0.1	0.9	−1.3	−1.2	−2.5
68	0.8	0.2	0.6	+0.4	+3.3	+3.7
11	0.5	0.3	0.2	+0.2	−0.1	+0.1
74	0.4	0.1	0.3	+0.2	−1.0	−0.8
69	0.4	0.2	0.2	−0.1	−3.0	−3.1
59	0.4	1.8	−1.4	+1.4	+0.9	+2.3
47	0.4	0.3	0.1	+0.2	−1.3	−1.1
40	0.4	0.2	0.2	−1.6	−8.8	−10.4
67	0.3	0.2	0.1	−1.9	+3.0	+1.1
5	0.1	0.2	−0.1	+0.4	+0.4	+0.8
1	0.1	0.2	−0.1	−1.5	−1.1	−2.6
10	0.1	−	0.1	−0.5	+0.6	+0.1
	151.0	35.2	115.8	−6.0	+23.1	+17.1

one might suppose that the excess of O_i over D_j traffic reflects the onward trans-shipment of imports to the interior of the country.

That this may not be the case in practice is suggested by data in table 3.6. There is some suggestion that those traffic zones that receive large volumes of imports are those for which the inland D_j traffic exceeds the O_i traffic by the greatest margin. However, simple linear regression, using the logarithm of import volume and, as an alternative, the excess of imports over exports as the independent variable, does not yield an acceptable correlation (even at the ninety-five per cent level) to explain the excess of D_j over O_i inland traffic. Despite this absence of an acceptable correlation, the slope of the regression

Freight flows and the British economy

Table 3.7. *Port zones: excess (+) or deficit (−) of D_j over O_i traffic, by mode and commodity group, million tons*

Commodity group	Road			Rail			Net difference
	O_i	D_j	Difference	O_i	D_j	Difference	
Coal and coke	63.9	65.6	+1.7	59.8	93.0	+33.2	+34.9
Iron ore	0	0	0	16.3	12.7	−3.6	−3.6
Limestone	0	0	0	1.2	2.0	+0.8	+0.8
Scrap	4.0	4.1	+0.1	2.9	2.5	−0.4	−0.3
Steel	16.5	16.4	−0.1	11.8	9.8	−2.0	−2.1
Oil	37.0	31.5	−5.3	6.3	3.3	−3.0	−8.3
Transport goods, etc.	3.6	4.0	+0.4	0.3	0.5	+0.2	+0.6
Food	191.2	184.2	−7.0	5.0	4.1	−0.9	−7.9
Chemicals	16.4	17.1	+0.7	2.7	3.2	+0.5	+1.2
Building materials	227.2	230.8	+3.6	7.7	8.1	+0.4	+4.0
Other manufactures	72.6	74.9	+2.3	1.2	1.2	0	+2.3
Other crude	40.9	37.9	−3.0	6.5	4.6	−1.9	−4.9
Miscellaneous	63.5	64.2	+0.7	0.7	0.5	−0.2	+0.5
Total	736.8	730.7	−6.1	122.4	145.5	+23.1	+17.0

Note: small differences in the totals compared with other tables arise from rounding; similarly, the body of the table does not exactly sum to the totals shown.

line is positive, indicating that so far as there is any tendency it is for the excess of D_j over O_i traffic to be greater where imports are absolutely or relatively more important. At the very least, therefore, we may conclude that the expected relationship does not exist and that there is a hint of the reverse being true.

Conclusion

On the basis of the evidence presented above, it is abundantly clear that the pattern of inland traffic flows of the port zones is very little affected by oversea trade. The implication is that in our analysis of the freight space-economy it will be reasonable to ignore the effects of port activities, at least in the main stages of the study. It might be relevant to examine the pattern of residuals to see whether foreign trade patterns could add to any explanation already achieved, but work done in the Department of the Environment indicates that even this would be unrewarding. Using the same basic data source that we have employed, the Department fitted a gravity model to inland road freight and then sought to 'improve' it by adding in a port variable to take account of the volume of imports and exports.

'The results showed that most port variables either were insignificant at the ninety-nine per cent level or had negative coefficients in the estimating equations obtained. The only port variable which was both significant and positive was exports of chemicals which, with a coefficient of 0.01, had almost no impact on the estimated flows' (Department of the Environment, 1971, 12).

In view of this corroboration of the evidence presented above, no further analysis of the port zones as a separate class has been undertaken in the present study, and we have treated Great Britain as though it were a closed economy.

Chapter 4

FREIGHT FLOW DATA

Until the last decade, no comprehensive data have been available for Great Britain on the geography of freight movements. The basic reason for this lack of information is that there is no statutory cause for any government agency to collect it on a regular basis, nor is there the occasion to collect data from which the geographical patterns would be extracted as a by-product. In this respect Britain differs from the United States, for example, where the Interstate Commerce Commission collects waybill data. Thus, the gap in our knowledge has only been filled by the mounting of special surveys, of which the first was in 1962.

1962 data

The 1962 survey of road freight traffic was based on a sample of 40,000 vehicles, drawn from a population of 400,000 vehicles in all. This population comprised all public haulage vehicles and the larger ones operated on an 'own account' basis, i.e., all 'C' licensed vehicles exceeding three tons unladen. The sample vehicles undertook approximately 700,000 journeys in the survey period, which was four separate weeks selected to represent the seasonal pattern; the sample vehicles were divided into four groups. A stratified random sample was used to get a suitable representation of variations by region (13 Ministry of Transport Traffic Areas), licence type (4 classes) and unladen weight (6 classes). Questionnaires were distributed to the transport operators, requesting full details of the work done by the specified vehicle(s) in the survey week. From the sample results, annual quantities were estimated (Ministry of Transport, 1966c).

Questions were asked on the origin and destination of journeys and the results were coded so that flows of goods in 15 classes (see table 4.1) between 107 zones could be produced. The 107 zones were subdivisions of Ministry of Transport Traffic Areas and aggregates of local authority areas. In compiling the matrices of freight flows, the Ministry dropped a distinction made in much of their analysis, i.e., the distinction between 'end-to-end' journeys and 'intermediate' ones. The latter are trips in which goods are picked up or set down at several points *en route*, whereas the former are journeys with a single origin and a single destination. Hence, for the origin/destination matrices, all intermediate traffic was counted as end-to-end, the destination being the point furthest from the origin. The effect of this convention is that the mean haul for intermediate traffic is inflated somewhat, a point to which we shall return.

30

Table 4.1. *1962 road freight survey commodity classes*

Code number	Commodity
1	Cereals
	Fresh fruits, vegetables, nuts and flowers
2	Meat and poultry
	Fish
	Live animals
	Dairy produce, eggs
3	Beverages
	Flour
	Animal feeding stuffs
	Other foods, tobacco
4	Oil seeds, nuts and kernels, animal and vegetable oils and fats
	Wood, timber and cork
	Crude and manufactured fertilisers
5	Crude minerals other than ore
6	Iron ore and scrap iron
	Non-ferrous metal ores
7	Textile fibres and waste
	Other crude materials
8	Coal and coke
9	Petroleum and petroleum products, gas
10	Tars from coal and natural gas
	Chemicals and plastic materials
11	Lime
	Cement
	Building materials
12	Iron and steel, finished and semi-finished products
	Non-ferrous metals
13	Metal manufactures
	Electrical and non-electrical machinery: transport equipment
14	Miscellaneous manufactured articles
15	Furniture removals
	Unallocable loads: mixed loads
	Empty containers
	Laundry and dry cleaning

Operators of light 'C' licensed vehicles (unladen weight three tons and less) were not asked to provide information on individual journeys. The total number of journeys and the tonnage carried were allocated arbitrarily to the intra-zonal category — the diagonal cells of the flow matrices — using the address at which the vehicle was garaged as the basis of allocation. This may somewhat overstate the volume of within-zone traffic relative to inter-zonal flows and will tend to produce an underestimate of mean journey length. We do not know whether this fully compensates for the exaggeration noted in the previous paragraph, though probably not, because 'C' licensed vehicles are mainly used for local delivery purposes.

Some information derived from this 1962 survey has been published (Ministry of Transport, 1964 and 1966b) and the results of some modelling exercises made available (Department of the Environment, 1971). The full

31

Freight flows and the British economy

tabulations of the 107 x 107 matrices have not been published, though they have been made available to researchers.

One important feature of the 1962 survey must be noted. The data were obtained from a survey of the operations of the road haulage industry. Thus, the origin and destination of goods means only the origin and destination for the particular movement: the same goods may appear again as a separate movement. For example, the goods moving into a wholesale warehouse will have that building as their destination; when a delivery is made thence, the warehouse will be recorded as the origin. Consequently, the freight flow data, while accurately representing the work done by the transport sector, do not faithfully reproduce the flows from first origin to final destination.

Finally, because the data have been obtained from a sample survey there are inevitable sampling errors. These are of course greatest for the smallest flows and least for the largest, though no formal estimate of their magnitude was made. Mr S. L. Edwards, who was responsible for much of the work on this survey and who is an excellent statistician, gave it as his opinion that the sampling errors were of the following magnitudes.

For a flow T_{ij}, in tons p.a.	Percentage error limits
10,000 tons	± 100%
100,000 tons	± 20%
1,000,000 tons	± 5%

These limits represent the probable ranges within which the true values lie, but are to be taken as no more than an informed estimate. The reader will appreciate, however, that sampling errors of the above magnitude dictate caution in handling the data. In particular, they rule out analyses in which small flows are an important element and have constrained us to work at a fairly aggregate level of analysis

1964 data

Within the Ministry of Transport, the Mathematical Advisory Unit (M.A.U.) were in 1968 developing a 'Transport Costs Model' (T.C.M.) for costing freight movement in this country (Ministry of Transport, 1968). For this purpose, compatible road and rail data were required. Some information for rail movements had previously been collected for 1964 (British Railways Board, 1965). Though no details are available on the way in which the data were obtained, it appears that wagon movements were sampled. This survey by British Railways was undertaken independently of the 1962 road survey and there were inevitable problems in matching the two sets of data.

In the first place, a compromise set of seventy-eight zones was used (see Appendix). These zones were mostly aggregations of the 107 zones but in some cases sub-division took place. To allocate the road freight was easy when zones

32

were amalgamated but posed problems where sub-divisions occurred. In the case of London, the 78-zone system provides a finer geographical breakdown than does the 107-zone one; but the London Traffic Survey allowed reasonably accurate allocation of the road freight flows. Elsewhere, road traffic was allocated to zones formed by sub-division of zones in the 107-cell system, in direct proportion to the resident populations of the parts.

The second problem of reconciliation concerned the commodities. As shown in table 4.2, the 15 commodity classes of the 1962 survey were compressed into

Table 4.2. *1964 commodity classes, derived from the 1962 classes*

Transport Costs Model commodity groups		Components from the 1962 Survey (table 4.1)
Number	Description	
1	Coal and coke	8
2	Iron ore	—
3	Limestone – for the steel industry	—
4	Iron and steel scrap	2/3 of 6
5	Steel (finished and semi-finished)	4/5 of 12
6	Oil	9
7	Transport vehicles and equipment	1/7 of 13
8	Foodstuffs	1, 2, 3
9	Chemicals	10
10	Building minerals and materials	5, 11
11	Other manufactures	14, 6/7 of 13, 1/5 of 12
12	Other crude materials	4, 7, 1/3 of 6
13	Miscellaneous	15

Note: iron ore and limestone move exclusively by rail and therefore did not appear in the 1962 survey of road freight.

thirteen. Two of these commodities – limestone and iron ore – move exclusively by rail, with the consequence that for road traffic there are only 11 commodity groups.

Once the road freight flows had been allocated to the seventy-eight zones and 13 commodities on a basis compatible with the rail freight data, it was necessary to gross up the 1962 volumes to 1964. This was done by applying a factor to each commodity matrix, as shown in table 4.3. These factors are assumed to be based on evidence regarding the growth of industries generating various traffics and also on evidence from the 1964 rail freight survey, but full details have not been published. Each factor was applied uniformly to its relevant commodity matrix, so that no change in the geographical pattern was introduced for the individual commodities. However, some change in the geographical pattern of the aggregate flows does arise, because of the change in the commodity structure of zones implied by the differential rates of growth applied to structures of freight traffic that vary from zone to zone.

For the purpose of the Transport Costs Model, the Mathematical Advisory Unit of the Ministry compiled a 78 x 78 matrix of road mileages, representing

33

Freight flows and the British economy

Table 4.3. *Factors for raising 1962 road tonnages to a 1964 equivalent, individual commodity groups*

Group 1	Coal and coke	Decreased by 2.5%
Group 4	Iron and steel scrap	Increased by 20.0%
Group 5	Steel (finished and semi-finished)	Increased by 20.0%
Group 6	Oil	Increased by 6.0%
Group 7	Transport vehicles and equipment	Increased by 13.0%
Group 8	Foodstuffs	Increased by 29.7%
Group 9	Chemicals	Increased by 21.0%
Group 10	Building minerals and materials	Increased by 17.5%
Group 11	Other manfactures	Increased by 57.0%
Group 12	Other crude materials	Increased by 8.5%
Group 13	Miscellaneous	Increased by 15.0%

the distances between and within zones. The inter-zonal distances are between the set of zone centroids, one centroid for each zone. In selecting the centroids, the Unit used the major traffic node or most important city in each zone. Thus, although the centroids are not necessarily related to the notion of the centre of gravity of activities within zones, they are relevant for the geographical pattern of transport flows and more particularly for estimating the need to modify existing links in the transport network. Intra-zonal distances were estimated as two-fifths of the square root of the area of the zone. This is an approximation to the moment representing the mean length of trip in a circular area with a uniform distribution of origins and destinations. As many zones are far from circular, this arbitrary method probably introduces a significant degree of error. Unfortunately, with the data in the form available to us, there is no better method of estimation.

It is the 1964 data that form the basis for the work reported in the present book. In only one case have the 1962 figures been used and this fact is clearly indicated at the appropriate point. From the brief account that has been given, the reader will realise that the freight flow data are imperfect. On the other hand, in the absence of even preliminary results from the 1967–8 survey of road freight also mounted by the Ministry, the 1962 and 1964 sets were literally the only comprehensive documentation of inland freight flows in this country at the time the work here reported was undertaken.

In the Appendix is a list of the zones, with the names of the cities used as centroids and also an outline map showing the location and extent of each zone. This Appendix also contains summary data for the zones, which the reader may wish to consult in elaboration of points made in the text.

Finally, it should be noted that the data used in the present study do not tally exactly with the published data for the aggregate tonnage and ton-mileage of freight. As table 4.4 shows, the tonnage estimates are very similar and the apparent discrepancies are attributable to rounding and to minor adjustments from the survey totals. But the differences in ton-mileage do not admit of this explanation, since the two estimates are arrived at by essentially different routes. The estimates published in the *Annual Abstract of Statistics* are based on an analysis of the survey data in their stratified form, and independently of any

34

Table 4.4. *Great Britain: comparison of freight volume estimates, 1964*

	Data used in the present study	Data published in the *Annual Abstract of Statistics*
Million tons		
Road	1,384	1,400
Rail	245	240
Total	1,629	1,640
Thousand-million ton-miles		
Road	45.9	39.0
Rail	14.7	16.1
Total	60.6	55.1

Source: Appendix and *Annual Abstract of Statistics,* 1970, p. 217.

allocation to zones of origin and destination. The ton-mileage data shown in the Appendix have been obtained from the tonnage-weighted mean haul, summed for inter- plus intra-zonal hauls. Given that the distance estimations used in manipulating the origin-destination matrices have defects, as noted above, it is not altogether surprising that discrepancies arise in the calculation of aggregate ton-mileage. However, though the discrepancies are significant, they are not of an order to prejudice the use of the 78 x 78 matrices.

Chapter 5

ESTIMATING THE GENERATION OF TRAFFIC

As a first step in modelling freight flows, it is necessary to find some basis on which to estimate the volume of goods that will be generated in each zone and attracted to it. Once a relationship between freight volume and one or more independent variables has been established, it becomes possible to predict the effects of changes in the independent variables upon the pattern of goods traffic. Ideally, the independent variables will themselves be amenable to accurate forecasting, in which case good forecasts can be made of the amount of freight that will be generated in and attracted to each zone. Even if the independent variables cannot be forecast accurately, it will be possible to undertake simulation exercises to examine the effects of possible future spatial dispositions of the independent variables.

Following the basic strategy of this book, it is desirable to start with the problem of forecasting the national aggregate volume of freight. The control total so obtained will then provide the basis on which allocations to the various zones can be made. In such a staged approach, we might expect the national aggregate to be more susceptible to forecasting than zonal totals; in practice, this proves not to be the case. Furthermore, it would be convenient if the national and zonal forecasts could be procedurally linked. In the event, this also is not possible.

National freight estimates

Surprisingly little work has been done in this country to relate growth in freight traffic to other indicators of economic growth. As Chisholm (1971c) has shown, most previous attempts have relied upon a combination of techniques: on the one hand, estimates of future traffic obtained from some basic industries like coal and steel; on the other, an assumed relationship between freight growth and changes in either the Gross Domestic or the Gross National Product. The forecasting relationship has, therefore, been a postulated rate of growth in freight volume linked to the growth of the economy. In some studies, it has been postulated that when the economy is growing slowly freight traffic increases fractionally less quickly, whereas if economic growth progresses at a respectable pace then the volume of goods to be moved will rise substantially faster. A major reason that has been offered for this variable relationship is that the building industry generates a large volume of traffic and that at times of slow growth this industry stagnates, while in good times it experiences something of a

36

boom. Apart from the fact that the year 1971 demonstrated the falseness of this particular proposition (economy stagnant, house purchasers scrambling for properties and civil construction continuing at a respectable level), it is clearly desirable to see whether a more formal forecasting relationship can be derived.

For the period 1953 to 1968, the national aggregates of freight tonnage and ton-miles can be compared with the Gross Domestic Product. To eliminate most of the temporal autocorrelation involved in using time series of this kind, regressions were calculated in terms of first differences. That is to say, the annual change in freight volume is compared with the annual change in G.D.P. The estimating equations so obtained for total tonnage yielded the following levels of correlation:

All freight	$R^2 = 0.29$
Road freight	$R^2 = 0.09$
Rail freight	$R^2 = 0.59$

None of these relationships is acceptable at the ninety-nine per cent level and that for road freight does not even achieve the ninety-five per cent level. This is a disappointing result because it means that changes in the future volume of freight cannot readily be predicted from changes in G.D.P. As a consequence, the analyst must resort to the rather crude technique of trend extrapolation, with all the hazards involved.

But of course G.D.P. or equivalent estimates are not available at the sub-national level, except for Scotland and Wales. At the level of disaggregation of the Ministry of Transport's seventy-eight zones, there is nothing available on an annual basis that remotely resembles national income data. Population and employment figures are, however, available annually at this level of disaggregation. Therefore it seems worth enquiring whether, at the national level, changes in freight volume per person could be related to time as the independent variable. With both dependent variables, freight volume related to population and employment, there is some association, but the relationship is so slight that it cannot be accepted even at the ninety-five per cent level.

These results lead to a disappointing conclusion. At the national level, there is a rather poor basis for estimating the future volume of freight that is to be moved and there is no ready means for linking such forecasts directly with the procedures for allocating freight to the seventy-eight zones. Essentially, therefore, forecasting at the national and at the zonal levels must be conceived as separate exercises. As a consequence, the most that we can do at present is to take the national freight total as a given quantity and ask whether efficient means can be found for allocating this total among the zones. As is shown in the next section, the answer is encouragingly affirmative, though the success achieved is not sufficient to warrant full confidence in the relationships established.

Zonal freight estimates

At the 78-zone level, there are three independent variables that can be regarded as surrogates for economic activity, and which ought to be related to freight volumes O_i and D_j, for which data are available. These are total resident

37

population, employed population and retail turnover. In this section, we report the results obtained in our attempts to explain the spatial distribution of freight tonnages attracted and generated using these three independent variables.

If we could assume no commuting across zonal boundaries, full employment and equal productivity and incomes, the total resident population would quite accurately represent the productive and consuming capacity of each zone. Even though these assumptions cannot be sustained, it is clearly worth experimenting with resident population as a surrogate for economic activity. For this purpose, we have used 1961 populations (in thousands) as recorded in the census, local authority areas being combined to provide zone totals. Though the freight flow data are for 1964, it was felt that using 1961 populations would not introduce any serious error and had the merit of being based on a full enumeration, whereas intercensal population totals are subject to errors of estimation. In any case, the Ministry of Transport subsequently compiled 1964 population totals and these correlate very highly with the 1961 figures ($R^2 = 0.96$).

The Ministry of Transport compiled estimates for total employment (in hundreds) for the year 1964 and also for employment in the twenty-four Orders of the Standard Industrial Classification. Employment is likely to be a closer surrogate for production than is resident population. Furthermore, employment in the S.I.C. Orders can be used as independent variables to provide a means of associating the production and attraction of commodities with the level of activity in industries which produce them as outputs, use them as inputs or handle them in trade.

Retail turnover provides the third independent variable available at the zonal level. It is a measure of final consumption and therefore may be related to the attraction of freight to zones. The data, in thousands of pounds sterling, were compiled from the 1961 Census of Distribution.

In this section, therefore, we examine the utility of these variables for forecasting freight volumes generated and attracted at the zonal level. Notwithstanding the dismal performance at the national level for changes over time, the independent variables perform reasonably well for inter-zonal allocation at a moment of time. As indicated in chapter 2, the tactic is to begin at the most aggregate level and then to disaggregate. Disaggregation is done first on the dependent side of the relationship, i.e., freight volume, and then on the independent side. The purpose is to seek the most appropriate level of aggregation for the job in hand, as judged by the levels of statistical 'explanation' and significance achieved and the reasonableness of the results. The technique used to establish the relationships is least squares regression – both simple and multiple. The tonnages of traffic originating and terminating in areas, in total, by different modes of transport and of different commodities, are considered to be positively and linearly related to the level of economic activity in the zones. It would seem sensible to surmise that as the level of activity in a zone (E_i) is greater so is the amount of traffic it originates (O_i) or attracts (D_j):

$$O_i = a_o + b_o E_i$$
$$D_j = a_d + b_d E_i,$$

38

where a is some constant, representing a threshold level of activity which must be achieved before the area starts generating traffic.

Having observations of the generation of traffic by zones and some indicators of the level of activity in the zones, we can establish the values of the parameters a and b using least squares, with the zones as observations. The dependent variable can be disaggregated according to the mode of transport by which the goods are carried (m) and by commodity classes (k) and the independent variable can be disaggregated into a number of sectors of economic activity $(E_1, E_2, \ldots E_n)$, so that we can fit expressions of the form:

$$O_{imk} = a_o + b_{o1}E_1 + b_{o2}E_2 + \ldots + b_{on}E_n,$$

$$D_{jmk} = a_d + b_{d1}E_1 + b_{d2}E_2 + \ldots + b_{dn}E_n,$$

using multiple regression.

Using least squares regression in this manner, we are assuming that traffic generation (or attraction) is a linear function of the level of economic activity in the zones. Starkie (1967) found evidence to show that the relationship is in fact curvilinear, i.e., that traffic does not increase as fast as the level of economic activity. In Starkie's study, however, the level of observation was individual, manufacturing plants and traffic was measured in terms of vehicle trips. The observed non-linear relationship can, therefore, be explained quite readily in terms of economies of scale. As the scale of a plant's activities increases, higher load factors can be achieved and larger vehicles used, so reducing the number of trips needed to move a given quantity of goods. These considerations do not apply in the present study, since we are working with freight tonnage and not vehicle trips. At the scale of analysis here employed, a non-linear relationship between freight volume and levels of economic activity would imply spatial variations in technology and hence of input-output relationships at the industry level, for which independent evidence is not available. In this context, note also that all freight flows are included in the analysis, both intra- and inter-zonal. Were the intra-zonal flows to be excluded, serious problems of scale would arise, because the proportion of intra-zonal to total traffic is greatly affected by the geographical extent of zones. Exclusion of the intra-zonal traffic would mean that freight volume (inter-zonal) would not vary directly with the level of economic activity, but inversely with the area of the zone.

All-in-all, therefore, it seems reasonable to assume that freight volume is a linear and positive function of the surrogates for economic activity. This assumption is confirmed by the fact that the residuals, or unexplained variation in freight volumes, display no kind of spatial association that could be interpreted in terms of regional variations in input-output relationships related to scale economies (see p. 59).

One difficulty arises because of the nature of the data we are using, which are derived from a survey of vehicle operations; origins and destinations represent the workings of the transport system rather than of production and consumption directly. Much traffic is being gathered to and redistributed from transport nodes and its movement reflects trade rather than production activities. These transport functions will obviously be reflected in employment and thus at the aggregate level this will not be too serious a problem. If we have these transport and trade activities represented among our activity variables at the zonal level, as is the case, then the problem should vanish. The amount of trade should be proportional to employment in these activities.

To summarise the independent variables available, the following brief listing is helpful:

> Total population, 1961, in thousands,
> Total retail turnover, 1961, in thousands of pounds sterling,
> Total employment, and by S.I.C. Order, 1964, in hundreds.

Procedure

The basic strategy adopted was to start from the most aggregate level of analysis and then progressively to disaggregate. Thus the total volume of freight, road plus rail, both originating from and terminating in the seventy-eight zones, was regressed on resident population and employed population. A similar regression was obtained for the terminating traffic, using retail turnover as the independent variable. The next step was to disaggregate the dependent variable by mode, i.e., into road and rail freight, and then again by commodity. After these analyses were completed, attention was turned to the disaggregation of the independent variable, employment. In this phase of the study, the aggregate commodity flow was regressed on employment in the S.I.C. Orders and then similar regressions were performed for commodity flows by road and rail separately.

With employment split up into twenty-four categories, a 'step-wise' multiple regression procedure with a cut-off at a conservative significance level was used. The 'step-wise' procedure successively adds independent variables to a regression equation in the order of the amount of the variance of the dependent variable that they statistically explain. As each new variable is picked up, statistical tests are carried out to establish the improvement made to the equation. A 'cut-off' in this procedure can be specified as the level of probability at which the regression coefficient of the next variable to be included is insignificantly different from zero according to the t or F tests. In this experiment, a very high probability level was specified: the coefficients included in the equation had to be significant at the ninety-nine per cent level according to the F test. Thus, only coefficients for which it was ninety-nine per cent certain that they were significantly different from zero were accepted. The 'step-wise' procedure used in this manner is very much a 'shot-gun' approach. Least squares computations in multidimensional multivariate space may produce significant negative coefficients which make no sense in terms of our initial hypothesis of positive relationships between traffic generation and production and consumption. The addition of each new variable in the procedure requires the recalculation of all the regression coefficients. This can result in what seem sensible and positive coefficients on employment in activities that one would intuitively associate with the generation of certain traffics, at one stage, becoming negative as further variables are included in the equation. To combat this problem, the step-wise procedure was retraced to the point where all coefficients were positive and significant at the ninety-nine per cent level. This rather laborious task was done for road and rail combined and not for the two modes separately.

Another problem in this exercise was the close relationship between many of the 'independent' variables. There are high correlations between employment in several industries over the zones. This might have been overcome by grouping the employment variables, either by inspection of a correlation matrix of all the independent variables (Department of the Environment, 1971), or by some more formal procedure in the form of a cluster analysis on the correlation matrix. An approach to the problem might be made by subjecting the correlation matrix to a principal components analysis which would produce a set of twenty-four synthetic, orthogonal components explaining all the variance of the original variables. It would then be possible to choose those employment variables which were loaded most highly on the first few components, which explain most of the total variance, as representing the major dimensions of employment variation over the country. But of course these variables would not necessarily be independent of one another. Rather than get involved in the time-consuming manipulations that are implied in these procedures, the arbitrary decisions which must necessarily be made and the difficulty of interpreting the results of a principal components analysis, we have preferred to use a simpler approach. In examining the outputs obtained from the step-wise procedures, we have been alert for results that appear unreasonable and for those which clearly involve highly intercorrelated 'independent' variables. This is not a foolproof procedure but can be justified by the purpose of this exercise. Our aim is not to establish rigorous causal inferences, but rather to achieve a high level of statistical 'explanation' that can be used for predictive purposes.

Results: gross variables

Table 5.1 shows the results obtained for total freight tonnage, and road and rail separately, regressed on the three independent variables of resident population, number employed and retail turnover. Only if the coefficient of determination (R^2) exceeds 0.08 is the correlation between the dependent variable and independent variable significant at the ninety-nine per cent level. The F values tabulated refer to the significance of the regression coefficients (b). With 76 degrees of freedom, an F value of 7.0 or over indicates that the regression coefficient of the estimating equation is significantly different from zero at the ninety-nine per cent level. Despite the non-significance of results for rail freight, the regressions for road plus rail account for a high proportion of the variance in the dependent variables (65 to 79 per cent) while for road alone the performance is rather better, ranging from 74 to 79 per cent. The poor results for rail traffic are not entirely surprising, since rail freight is dominated by mineral and bulk goods traffics, which emanate from a limited number of origins and are consigned to a few major points of consumption; for both origins and destinations there is no very close association with the gross

41

Freight flows and the British economy

Table 5.1. *Total freight tonnage generated, regressed on population, employment and retail turnover*

Dependent variable	Independent variable	Constant a	Coefficient b	F value (b coefficient)	Correlation R^2
Total road and rail tonnage					
O_i	Population (000)	40,867	256.06	216.7	0.74
D_j	Population (000)	38,339	262.08	278.0	0.79
O_i	Employment (00)	65,275	4.70	146.1	0.66
D_j	Employment (00)	60,593	4.81	175.3	0.70
D_j	Retail turnover (£000)	58,237	1.42	144.3	0.65
Total road tonnage					
O_i	Population (000)	27,576	222.13	254.2	0.77
D_j	Population (000)	25,657	227.09	289.3	0.79
O_i	Employment (00)	42,178	4.39	252.8	0.77
D_j	Employment (00)	41,856	4.43	263.8	0.78
D_j	Retail turnover (£000)	39,129	1.31	210.9	0.74
Total rail tonnage					
O_i	Population (000)	15,376	24.83	6.5	0.08
D_j	Population (000)	14,564	26.08	11.0	0.13
O_i	Employment (00)	20,397	0.36	3.5	0.04
D_j	Employment (00)	1,837	0.42	7.0	0.08
D_j	Retail turnover (£000)	1,251	0.13	4.2	0.05

independent variables used. Perhaps of more interest is the fact that retail turnover performs less well than either population or employment in accounting for the volume of attracted (D_j) traffic, and that resident population is slightly more useful than employed population. Also noteworthy is the fact that a slightly better explanation is achieved for the terminating (D_j) than for the originating (O_i) traffic, though the difference is small.

When we disaggregate the freight volumes by commodity, we get the results set out in table 5.2. As previously, rail traffic yields rather poor and mostly non-significant results and for this mode the only commodity for which R^2 rises above 0.50 is other manufactures, a heterogeneous class that contributes only a small part to the total rail traffic. On the other hand, road freight yields significant results. The median values for R^2 for this mode are:

	Independent variable	
	Population	Employment
Originating (O_i)	0.48	0.50
Terminating (D_j)	0.49	0.52

42

Table 5.2. *Tonnages of commodities regressed on population and employment*

Independent variable

Dependent variable	Population (000)				Employment (00)			
	a	b	F (b coefficient)	R^2	a	b	F (b coefficient)	R^2
Tonnage of commodities:								
Road and rail								
O_i: Coal and								
coke	16,671	34.48	9.3	0.11	2,214	0.55	6.0	0.07
Iron ore	1,013	2.37	1.7	0.02	1,391	0.03	1.1	0.01
Limestone	400	−0.10	0.1	0.00	494	0.00	0.2	0.00
Scrap	87	1.97	30.5	0.27	650	0.05	26.8	0.26
Steel	2,544	11.51	11.1	0.13	3,666	0.21	9.2	0.10
Oil	804	11.74	47.4	0.39	1,241	0.23	51.0	0.40
Transport goods, etc.	−100	1.88	115.0	0.60	14	0.04	105.2	0.58
Food	8,122	49.18	108.3	0.59	9,649	0.98	127.3	0.63
Chemicals	2,074	4.60	28.6	0.27	2,394	0.08	26.2	0.26
Building materials	11,222	73.67	144.9	0.66	21,055	1.21	72.1	0.49
Other manufactures	−482	30.70	186.3	0.71	512	0.62	232.3	0.75
Other crude	2,438	11.35	71.8	0.49	3,343	0.21	59.6	0.44
Miscellaneous	−154	22.71	282.3	0.79	893	0.44	302.4	0.80
D_j: Coal and								
coke	15,237	34.72	25.9	0.25	19,889	0.57	17.6	0.19
Iron ore	1,825	1.10	0.6	0.01	2,146	0.01	0.2	0.00
Limestone	460	−0.07	0.1	0.00	462	0.00	0.0	0.00
Scrap	611	2.43	13.8	0.15	602	0.08	6.8	0.01
Steel	19	12.35	25.8	0.25	602	0.24	26.2	0.25
Oil	−273	13.37	156.8	0.67	714	0.25	125.0	0.62
Transport goods, etc.	17	1.72	136.3	0.64	88	0.03	148.8	0.66
Food	12,759	39.61	58.7	0.44	12,882	0.83	78.2	0.51
Chemicals	1,478	5.52	73.3	0.49	1,855	0.10	65.2	0.46
Building materials	9,431	76.42	207.2	0.73	19,633	1.28	90.3	0.54
Other manufactures	259	29.51	163.2	0.68	448	0.61	279.9	0.79
Other crude	2,569	12.69	52.2	0.41	3,886	0.22	38.8	0.34
Miscellaneous	186	22.52	313.8	0.81	1,563	0.43	263.9	0.78
Road								
O_i: Coal and								
coke	10,347	16.04	21.1	0.22	7,651	0.34	33.4	0.31
Scrap	151	1.97	27.2	0.26	144	0.04	34.7	0.31
Steel	1,014	10.45	15.3	0.17	1,829	0.19	13.7	0.15

Table 5.2. (cont.)

	Independent variable							
	Population (000)				Employment (00)			
Dependent variable	a	b	F (b coefficient)	R^2	a	b	F (b coefficient)	R^2
Road								
O_i (cont.)								
Oil	437	10.71	61.1	0.45	7,108	0.22	74.8	0.50
Transport goods, etc.	−99	1.73	112.3	0.60	−4	0.03	108.1	0.59
Food	7,964	48.20	111.7	0.60	9,415	0.97	132.7	0.64
Chemicals	2,892	3.79	9.5	0.11	1,765	0.08	40.9	0.35
Building materials	10,002	71.76	155.1	0.67	19,775	1.20	77.1	0.50
Other manufactures	−477	30.33	181.8	0.71	466	0.61	229.4	0.75
Other crude	2,086	10.46	70.9	0.48	2,826	0.20	62.3	0.45
Miscellaneous	−159	22.54	284.7	0.79	942	0.44	291.6	0.79
D_j: Coal and coke	5,520	19.51	66.0	0.47	6,667	0.31	40.1	0.35
Scrap	88	2.06	32.1	0.30	201	0.05	7.2	0.01
Steel	−489	9.72	27.8	0.27	201	0.14	26.5	0.26
Oil	−281	11.82	168.6	0.69	6,085	0.27	135.2	0.65
Transport good, etc.	4	1.59	122.5	0.62	50	0.03	150.7	0.67
Food	12,663	35.53	57.5	0.43	1,175	0.93	85.1	0.52
Chemicals	1,138	5.12	72.1	0.49	1,234	0.08	66.3	0.47
Building materials	8,629	74.28	217.1	0.74	18,425	1.32	95.1	0.55
Other manufactures	200	30.00	200.1	0.70	408	0.57	281.5	0.79
Other crude	2,336	10.53	64.0	0.46	3,375	0.18	41.3	0.35
Miscellaneous	175	22.37	306.6	0.80	1,574	0.42	274.6	0.78
Rail								
O_i: Coal and coke	10,938	15.32	3.0	0.04	12,885	0.22	2.0	0.03
Iron ore	1,013	2.37	1.7	0.02	1,391	0.04	1.1	0.01
Limestone	461	−0.10	0.1	0.00	494	0.00	0.2	0.00
Scrap	488	0.00	0.5	0.00	505	0.01	3.2	0.04
Steel	1,654	0.00	0.8	0.01	1,837	0.00	0.3	0.00
Oil	365	1.00	3.9	0.05	603	0.01	1.8	0.02
Transport goods, etc.	−2	0.15	16.7	0.18	19	0.00	11.9	0.13
Food	159	0.98	15.5	0.17	234	0.02	13.9	0.15
Chemicals	603	0.00	0.0	0.00	629	0.00	0.0	0.00
Building materials	856	2.04	7.8	0.09	1,849	0.00	0.0	0.00

Table 5.2. (cont.)

Independent variable

Dependent variable	Population (000)				Employment (00)			
	a	b	F (b coefficient)	R^2	a	b	F (b coefficient)	R^2
Rail								
O_i (cont.)								
Other manufactures	−6	0.36	85.6	0.53	46	0.01	47.6	0.39
Other crude	353	0.99	5.7	0.07	517	0.01	3.3	0.04
Miscellaneous	6	0.16	12.4	0.14	−49	0.01	45.8	0.38
D_j: Coal and coke	9,718	15.21	8.4	0.10	12,224	0.23	5.3	0.06
Iron ore	1,825	1.10	0.6	0.01	2,146	0.01	0.2	0.00
Limestone	461	−0.07	0.1	0.00	462	0.00	0.0	0.00
Scrap	522	0.40	0.8	0.01	582	0.01	0.5	0.00
Steel	508	2.57	8.6	0.10	857	0.04	6.2	0.08
Oil	8	1.58	25.2	0.25	107	0.03	23.8	0.24
Transport goods, etc.	13	0.13	20.1	0.21	106,700	3.86	0.0	0.00
Food	95	1.07	50.2	0.40	209	0.02	37.1	0.33
Chemicals	340	0.39	11.0	0.13	405	0.01	6.8	0.08
Building materials	803	2.13	17.3	0.19	1,218	0.03	9.1	0.10
Other manufactures	39	0.32	84.1	0.53	68	0.01	61.9	0.45
Other crude	223	1.06	11.8	0.13	388	0.01	7.8	0.09
Miscellaneous	11	0.15	14.5	0.16	10	0.02	13.2	0.15

The commodity groups with values of R^2 generally above the median values shown on p. 42 are: transport goods and equipment, food, building materials, other manufactures and miscellaneous, with the addition of oil in the case of the D_j traffic. For road and rail combined, the proportion of the variance accounted for is respectable, but generally less than in the case of road freight alone. Overall, it is the mineral traffic, with the partial exception of oil, for which these regression equations perform least well. The more general categories, especially building materials, other manufactures and miscellaneous, are very well accounted for by either population or employment as the independent variable. Finally, it is not obvious that either of the two independent variables performs better than the other, except in the case of road traffic O_i; in this case, employment seems in general to produce higher levels of R^2 than does population.

At this level of aggregation, we are also able to regress the commodity volumes attracted on retail turnover as the independent variable, with the

45

Table 5.3. *Tonnages of commodities attracted regressed on retail turnover*

Dependent variable	Independent variable			
Tonnage of	Retail turnover (£000)			
commodities attracted (D_j):	a	b	F (b coefficient)	R^2
Road and rail				
Coal and coke	20,520	0.16	14.4	0.16
Iron ore	2,267	0.00	0.1	0.00
Limestone	500	0.00	0.3	0.00
Scrap	917	0.01	9.1	0.11
Steel	915	0.06	20.7	0.21
Oil	602	0.07	106.1	0.58
Transport goods, etc.	73	0.01	123.5	0.62
Food	13,101	0.24	63.1	0.45
Chemicals	1,898	0.03	52.6	0.41
Building materials	18,057	0.38	87.2	0.53
Other manufactures	402	0.18	194.3	0.72
Other crude	3,446	0.07	40.5	0.35
Miscellaneous	1,010	0.13	249.4	0.77
Road				
Coal and coke	7,868	0.09	37.4	0.33
Scrap	285	0.01	22.8	0.23
Steel	21	0.05	24.2	0.24
Oil	501	0.06	111.5	0.59
Transport goods, etc.	42	0.01	119.9	0.61
Food	12,975	0.23	62.0	0.45
Chemicals	1,512	0.03	52.9	0.41
Building materials	16,956	0.38	90.1	0.54
Other manufactures	300	0.20	200.2	0.70
Other crude	3,073	0.06	36.5	0.32
Miscellaneous	1,002	0.13	243.3	0.76
Rail				
Coal and coke	12,651	0.06	4.2	0.05
Iron ore	2,267	0.00	0.1	0.00
Limestone	500	0.00	0.3	0.00
Scrap	632	0.00	0.2	0.00
Steel	1,143,400	0.16	0.0	0.00
Oil	101	0.01	21.4	0.22
Transport goods, etc.	31	0.00	12.3	0.14
Food	126	0.01	48.4	0.39
Chemicals	386	0.00	7.6	0.09
Building materials	1,101	0.01	10.5	0.12
Other manufactures	41	0.001	93.1	0.55
Other crude	373	0.01	7.5	0.09
Miscellaneous	8	0.00	16.7	0.18

results shown in table 5.3. The picture is essentially the same as that presented in table 5.2 and no special comment is called for.

The different independent variables used so far are highly correlated, as is shown by the following matrix of values for R:

	Population	Employment	Retail turnover
Population	1.00	0.87	0.92
Employment	0.87	1.00	0.94
Retail turnover	0.92	0.94	1.00

It is therefore not surprising that the results obtained are essentially similar. Though there is some suggestion in table 5.1 that population performs better than employment, this hint is not substantiated in table 5.2.

For the purpose in hand, it is rather serious that the explained proportion of variance in the dependent variable exceeds seventy per cent in only a few cases; in only one or two cases, and then mainly for the miscellaneous goods, is eighty per cent achieved. The implication is clear: the variables so far examined provide an inadequate basis on which to allocate to zones the relevant volume of freight, either in aggregate or on a commodity-by-commodity basis. Our attention must, therefore, turn to the disaggregation of the independent variable, employment, into the twenty-four S.I.C. Orders.

Results: employment by S.I.C. Orders

As an aid to understanding the discussion in this section, and also the results shown in tables 5.4, 5.5, 5.6 and 5.7, it is helpful to list the twenty-four S.I.C. employment Orders and the numbers that have been assigned to them. These numbers appear as subscripts to the X variables denoting employment in the specified S.I.C. Order. For example, X_1 means employment in Order 1, agriculture, forestry and fishing. The Orders are:

(1) Agriculture, forestry and fishing
(2) Mining and quarrying
(3) Food, drink and tobacco
(4) Chemicals
(5) Metal manufacture
(6) Engineering and electrical goods
(7) Shipbuilding and marine engineering
(8) Vehicles
(9) Metals not elsewhere specified
(10) Textiles
(11) Leather goods
(12) Clothing and footwear
(13) Bricks, pottery, glass and cement
(14) Timber and furniture
(15) Paper, printing and publishing
(16) Other manufactures
(17) Construction
(18) Gas, electricity and water
(19) Transport and communications
(20) Distributive trades
(21) Insurance, banking and finance
(22) Professional and scientific services
(23) Miscellaneous
(24) Public administration and defence

Tables 5.4, 5.5 and 5.6 display the results of the step-wise regression analysis of commodity tonnages generated by and attracted to the seventy-eight zones,

47

Freight flows and the British economy

for road plus rail, road only and rail only respectively. In all three tables, coefficients acceptable at the ninety-nine per cent level are included, irrespective of their sign. Table 5.4 shows a very respectable level of explanation for many commodities, at 80–90 per cent. Limestone is the commodity with the least variance explained, both for generated and attracted freight, followed by iron ore. Both these commodities move exclusively by rail and both generate a relatively small amount of employment, even as a proportion of the X_2 variable, mining and quarrying. Apart from these two commodities, the only squares of multiple correlation that fall below seventy per cent are the generation of oil and the attraction of scrap: the former commodity, though widely used, has a limited number of origin points at refineries and distribution depots; the latter is produced widely but used at only a few locations, not including all steel plants. Altogether, though, the results in Table 5.4 are most encouraging.

If we compare table 5.4 with table 5.5, it is evident that better explanations are achieved for road freight alone than for road plus rail, though there is a dramatic worsening in the case of the attraction of chemicals. Steel and building materials emerge with the best results, in both cases substantially over ninety per cent for inward as well as outward traffic. Turning to table 5.6, we see that for some commodities very reasonable levels of explanation are achieved for rail freight – notably for coal, scrap, steel, transport equipment, food and other manufactures. Very poor results are shown for limestone and building materials. While it is true that the estimating equations perform less well for rail traffic than for road, they nevertheless yield significant results at a good level of explanation. This is in marked contrast with the situation revealed in tables 5.2 and 5.3, where the majority of the results for rail are non-significant.

Disaggregation of the independent variable, employment, into the twenty-four S.I.C. Orders substantially improves the performance of the regressions. But, as presented in tables 5.4, 5.5 and 5.6, the results do contain a specious element, in the form of negative b-coefficients. Given our initial postulate of a positive and linear relationship between employment and freight volume, it is not very useful to dwell in detail on the significant independent variables so far identified. As mentioned previously, the next step is to suppress the negative coefficients, as in table 5.7.

For this final stage of analysis, we have chosen to work with road plus rail. The main reason for this choice is that road plus rail is the most aggregate data set we have and it performs nearly as well as road alone and much better than rail. In the following paragraphs, we comment on the commodity equations in table 5.7.

Coal and coke. Employment in mining and quarrying (X_2) is the main variable for both generation and attraction, a plausible result given that coal mining dominates this Order and a great deal of coal is used in or very near the coalfields. The inclusion of gas, electricity and water (X_{18}) in the equation for traffic generation is also sensible, since in 1964 coke was still a significant by-product from the manufacture of gas. The inclusion of the other variables is

48

Dependent variable	a	b1 / b13	b2 / b14	b3 / b15	b4 / b16	b5 / b17	b6 / b18	b7 / b19	b8 / b20	b9 / b21	b10 / b22	b11 / b23	b12 / b24	R^2
Road and rail tonnage of goods generated by zones (O_i)														
Coal and coke	−979.7		36.4		−13.6	5.1	19.3				6.2			0.90
Iron ore	−355.3					3.9		3.9		−3.7 / −5.8	−21.8	49.6	8.2	0.62
Limestone	154.5		0.4								0.3			0.18
Scrap	351.5	−0.1	−0.2 / −0.9	1.1		1.5								0.87
Steel	126.0			−3.6		14.9		2.6		−5.5				0.85
Oil	1,020.7				4.8	5.5	−1.3	2.8		2.5		4.6		0.63
Transport goods, etc.	−23.5			0.3	0.3	1.0	0.4		3.0	0.2			2.7	0.92
Food	3,202.2	9.6		26.1	5.6			3.3			−8.1 / 0.7	5.4		0.89
Chemicals	206.8		−8.8	1.4 / −1.5	6.3			1.2	0.5	0.7				0.81
Building materials	3,619.0	16.5 / 9.2	5.9 / −43.3	−14.6	9.6			20.9	−16.6	9.3 / −23.4	4.2 / 26.1		10.3	0.93
Other manufactures	632.9			7.3				6.4		6.2 / −7.9	2.2			0.90
Other crude	847.8	1.9		5.9							1.5			0.71
Miscellaneous	−55.0		−1.2		−8.1	2.2		5.0		3.2 / 4.0	2.0		1.4	0.89

Note: the X variables are defined on p. 47.

Table 5.4. (cont.)

Dependent variable	a	b1 / b13	b2 / b14	b3 / b15	b4 / b16	b5 / b17	b6 / b18	b7 / b19	b8 / b20	b9 / b21	b10 / b22	b11 / b23	b12 / b24	R^2
Attracted by zones (D_j):														
Coal and coke	2,114.5	5.0	8.7	-9.1	17.5	18.0		6.7		-7.6	7.8	53.9		0.85
Iron ore	158.4				1.5	5.8				-4.0				0.59
Limestone	153.9		-0.2		-0.8	0.7				-0.5 / -0.1				0.43
Scrap	-1,041.5	1.4 / 2.5				2.2								0.32
Steel	34.5	-0.7 / -6.7	-1.1	2.0		6.5			1.0	1.6				0.97
Oil	917.5	0.7 / -8.0		2.8 / 4.1	2.8 / 1.2				1.3	1.5				0.89
Transport goods, etc.	1.3							0.5	0.2	0.3 / -0.4			0.1	0.92
Food	6,608.7	9.0		29.6				6.7			-5.9			0.81
Chemicals	489.0	-6.0		1.2	2.9	0.3		1.8		0.7 / -1.9	0.3	9.3		0.87
Building materials	-316.0	17.5 / 8.2	4.7 / -47.1	-12.3	13.2		3.5	18.6	-6.7	7.3 / -13.6	5.5 / 20.0	-11.4	7.0	0.94
Other manufactures	581.6						1.9	5.9	1.2	4.0 / -2.3	1.9			0.90
Other crude	364.5			4.3 / 7.8	9.0	-2.1	3.7 / -25.2	3.4	2.6 / 5.3	2.0 / -11.6				0.81
Miscellaneous	635.0		-1.0	3.9		4.8	-1.3			2.2	1.8			0.86

Table 5.5. Road traffic by commodities regressed on employment in 24 S.I.C. Orders

Dependent variable	a	b1 / b13	b2 / b14	b3 / b15	b4 / b16	b5 / b17	b6 / b18	b7 / b19	b8 / b20	b9 / b21	b10 / b22	b11 / b23	b12 / b24	R^2
Tonnage of goods generated by zones (O_i):														
Coal and coke	2,568.5		7.7 / −1.0			2.2	1.8				3.6			0.85
Scrap	139.3		−0.2 / −1.0	0.7		0.4				0.7	0.3			0.87
Steel	79.2			1.1	−1.9	2.8		1.3		3.8	0.2			0.97
Oil	616.0	0.9		3.1	4.4					1.8				0.65
Transport goods, etc.	20.4	−0.1		0.2		0.3			0.2	0.3	−0.1		0.1	0.92
Food	3,337.4	9.1		24.7	5.9			3.2			−7.8	5.3		0.89
Chemicals	280.5		−7.6	1.2	3.9			1.4		1.0 / −1.0	0.5			0.81
Building materials	1,876.0	15.4 / 8.9	5.4 / −34.9	−9.7	6.7	21.6		12.9	−13.4	6.0	5.2 / 19.9	−13.3		0.92
Other manufactures	81.7			3.8		3.3	3.0	1.3				22.8		0.87
Other crude	871.7	1.9	−0.8	5.3							1.7			0.76
Miscellaneous	482.6			5.6	−4.6		9.5			2.4	1.2		1.4	0.87

Note: the X variables are defined on p. 47.

Table 5.5. (cont.)

Dependent variable	a	b1 / b13	b2 / b14	b3 / b15	b4 / b16	b5 / b17	b6 / b18	b7 / b19	b8 / b20	b9 / b21	b10 / b22	b11 / b23	b12 / b24	R²
Tonnage of goods attracted by zones (D_j):														
Coal and coke	900.9		4.0	−6.0		3.1 / 3.8	−19.3		3.3		4.0			0.89
Scrap	63.1			0.9	−1.4	0.4				0.7	0.3			0.90
Steel	52.6			3.7		4.9		1.5		1.9	0.3			0.96
Oil	971.0	0.8	−4.4		2.1 / 5.6	1.3			0.9					0.89
Transport goods etc.	107.6						0.1	0.1	0.2		0.2			0.87
Food	4,102.4	9.5		19.0				3.8						0.88
Chemicals	168.4	0.1 / −0.5		0.4	0.3	0.3	−0.1			−0.2				0.51
Building materials	602.5	16.8 / 7.2	4.9 / −43.8	−11.1	11.4		3.7	17.6	−6.7	6.7 / −12.5	5.2 / 19.8	11.3	6.7	0.94
Other manufactures	551.6						1.0	5.7	1.2	4.0 / −2.1	1.8			0.90
Other crude	789.6	2.3	−0.6	4.1						0.6	2.0			0.79
Miscellaneous	124.9		−0.9	4.1		4.6	−1.2			2.2	1.7			0.87

Table 5.6. *Rail traffic by commodities regressed on employment in 24 S.I.C. Orders*

Dependent variable	a	b1/b13	b2/b14	b3/b15	b4/b16	b5/b17	b6/b18	b7/b19	b8/b20	b9/b21	b10/b22	b11/b23	b12/b24	R^2
Rail tonnage of goods generated by zones (O_i):														
Coal and coke	-3,820.3		25.5			6.0				-4.3	3.9			0.93
Iron ore	-355.3					3.9		3.9		-3.8 -5.8	-21.8	49.4	8.2	0.62
Limestone	154.5		0.4			1.1								0.18
Scrap	178.6	-0.1		0.3		1.1				0.7				0.78
Steel	-522.8				2.4	5.4				-3.9				0.71
Oil	-312.9			-1.7	1.6	1.6		0.9				-1.7		0.39
Transport goods etc.	20.0			0.1		1.6		0.1	2.0	0.2			1.1	0.70
Food	-174.4	0.2		1.6		0.1						-2.7		0.69
Chemicals	110.0		-1.3		2.2 -0.9		-0.2							0.59
Building materials	1,365.3	0.6					0.6							0.07
Other manufactures	2.6		-0.2				-0.6			-0.1	0.1	0.1	-0.1	0.76
Other crude	71.5	-0.6			3.3	0.8		0.6	0.2	-0.7 -1.0				0.50
Miscellaneous	14.7	-0.1			-0.1								0.1	0.74

Note: the X variables are defined on p. 47.

Table 5.6. (cont.)

Dependent variable Rail tonnage of goods attracted by zones (D_j):	a	b1/b13	b2/b14	b3/b15	b4/b16	b5/b17	b6/b18	b7/b19	b8/b20	b9/b21	b10/b22	b11/b23	b12/b24	R^2
Coal and coke	−2,132.4	3.3 / −16.5	4.8	16.0	16.6 / 11.6					−12.1	2.9			0.72
Iron ore	158.4				1.5	5.8				−4.0				0.59
Limestone	153.9		−0.2		0.8	0.7				−0.5				0.43
Scrap	218.4		−0.2			2.1			−0.1	−1.1 / 0.4		−5.8		0.82
Steel	28.1	−0.3 / −1.0	−3.5		3.5	3.9	2.8			−2.2				0.79
Oil	−268.9					1.0			0.5	−0.4		−3.7 / 0.2		0.72
Transport goods etc.	−15.2		−0.2		0.2			0.2		−0.1				0.64
Food	3.0	0.3		0.8		−0.3	−0.1 / 1.9			0.2		−2.5		0.80
Chemicals	168.4	0.1	−0.5	0.4	0.3	0.3	−0.1			−0.2				0.51
Building materials	182.2	0.8	−3.6	−1.4	1.9						0.9			0.40
Other manufactures	18.2			−0.2					0.1					0.72
Other crude	169.3				1.1	0.4 / 0.8	−0.4					−0.2		0.32
Miscellaneous	49.0	−0.2	−0.2	0.2			0.6		−0.1	−0.1				0.38

Table 5.7. Revision of table 5.4 with only positive coefficients

Dependent variable	Estimating equation: Y is freight in 00 tons, X is employment in 00's	R^2
Road and rail tonnage of commodities generated by zones (O_i):		
Coal and coke	$Y = -1{,}514.9 + 37.0X_2 + 4.0X_5 + 6.5X_{10} + 9.6X_{18}$	0.90
Iron ore	$Y = 770.7 + 2.7X_{12}$	0.10
Limestone	$Y = 154.5 + 0.4X_2$	0.18
Scrap	$Y = 183.9 + 0.7X_3 + 1.5X_5 + 0.3X_{10}$	0.85
Steel	$Y = 2{,}227.5 + 10.4X_5$	0.76
Oil	$Y = -218.3 + 5.3X_4 + 1.5X_9 + 1.7X_{17}$	0.53
Transport goods, etc.	$Y = -23.5 + 0.3X_3 + 0.1X_5 + 0.3X_8 + 0.2X_9 + 0.3X_{16} + 0.4X_{18}$	0.92
Food	$Y = 3{,}295.7 + 8.8X_1 + 21.2X_3 + 3.9X_{19}$	0.87
Chemicals	$Y = 873.6 + 5.5X_4 + 1.0X_5$	0.65
Building materials	$Y = 6{,}295.0 + 14.0X_1 + 11.0X_2 + 13.0X_{17}$	0.82
Other manufactures	$Y = 222.7 + 3.9X_3 + 1.5X_5 + 1.8X_6 + 3.4X_9 + 1.9X_{10} + 2.5X_{19}$	0.90
Other crude materials	$Y = 874.8 + 2.0X_1 + 6.0X_3 + 1.5X_{10}$	0.71
Miscellaneous	$Y = 346.0 + 5.1X_3 + 2.0X_9 + 1.3X_{10} + 7.0X_{18} + 1.7X_{24}$	0.87
Road and rail tonnage of commodities attracted by zones (D_j):		
Coal and coke	$Y = 4{,}342.8 + 13.3X_2 + 18.6X_4 + 9.7X_5 + 5.3X_{10}$	0.75
Iron ore	$Y = 888.9 + 2.2X_5$	0.25
Limestone	$Y = 220.5 + 0.3X_5$	0.12
Scrap	$Y = -1{,}041.6 + 1.4X_1 + 2.2X_5 + 2.5X_{13}$	0.32
Steel	$Y = 221.1 + 1.5X_3 + 6.9X_5 + 1.7X_9$	0.96
Oil	$Y = -1{,}012.7 + 1.1X_1 + 2.1X_3 + 1.9X_4 + 1.6X_5 + 1.5X_8 + 5.5X_{16}$	0.87
Transport goods, etc.	$Y = 89.5 + 0.2X_3 + 0.1X_6 + 0.2X_8 + 0.2X_9 + 0.1X_{19}$	0.89
Food	$Y = 5{,}954.7 + 5.8X_1 + 28.4X_3$	0.76
Chemicals	$Y = 737.2 + 1.7X_3 + 2.9X_4 + 1.2X_5$	0.75
Building materials	$Y = 12{,}317.7 + 19.55X_{17}$	0.69
Other manufactures	$Y = 2{,}257.1 + 5.2X_9 + 1.8X_{10} + 2.2X_{19} + 1.7X_{20}$	0.88
Other crude materials	$Y = 73.4 + 8.1X_3 + 6.4X_7 + 1.7X_{10}$	0.60
Miscellaneous	$Y = -79.0 + 4.2X_3 + 1.4X_9 + 1.3X_{10} + 3.6X_{17}$	0.85

Note: the X variables are defined on p. 47.

probably due to the spatial association of chemicals (X_4), metal manufacturing (X_5) and textiles (X_{10}) with coal mining, rather than to industrial linkage, except that steel smelting is a large user of coal. Since in the step-wise procedure variables X_5 and X_{10} appear before X_{18}, it would not be logical in the present analytical framework to suppress the two former while retaining the latter. Nevertheless, for the generation of coal and coke freight, the variable X_2 accounts for eighty-four per cent of the variance:

$$Y = 9,519.6 + 41.3\ X_2,$$

where Y is the volume of freight in hundreds of tons and X_2 is employment, also in hundreds.

Iron ore. This commodity, which moves exclusively by rail, cannot be adequately accounted for on the generation side. The only significant relationship it has is with clothing and footwear (X_{12}), an industry which clearly has no important functional linkages with the mining of iron ore. It may be that the association is due to the concentration of footwear and clothing manufacture in and around Northampton, which is near the major domestic ore fields. On the attraction side, the appearance of metal manufacture (X_5) is plausible but does not result in a high level of explanation ($R^2 = 0.25$).

Limestone. Mining and quarrying (X_2) is reasonable as a variable to explain the generation of freight, as is metal manufacture (X_5) to account for its destination. But for this rail-borne commodity the level of explanation achieved is poor.

Scrap. Though eighty-five per cent of the production of scrap tonnage is accounted for by three variables, it is difficult to find a reason for textiles (X_{10}) being included. Metal manufacture (X_5) is clearly reasonable, since a good deal of scrap is produced as a by-product of basic metal working processes. Inclusion of food, drink and tobacco (X_3) may seem odd. However, this is a very widely dispersed industry, characteristic of all major urban areas; therefore it probably stands proxy for the fact that scrap is produced from the use of vehicles and other durable goods by the population at large. As for the attraction of scrap, metal manufacture (X_5) makes real sense as an explanatory variable. but the other two variables do not seem very plausible.

Steel. A clear and simple relationship between metal manufacture (X_5) and the generation of steel traffic is a gratifying result that makes good sense. A little odd, perhaps, is the fact that this same variable is important in explaining the destination of steel traffic, rather than Orders 6, 7 or 8. However, the inclusion of metals not elsewhere specified (X_9) is clearly acceptable, while food, drink and tobacco (X_3) is presumably to be regarded as a proxy for the general distribution of population, and is also a major user of steel (cans).

Oil. Most oil originates as inland traffic either from refineries or distribution depots; in both cases, therefore, from a coastal location. Consequently, it is not surprising that the generation of oil freight is associated with chemicals (X_4) and metals not elsewhere specified (X_9), since both of these industries are located in or near dockside areas. There is no evident reason for the inclusion of construction (X_{17}). The level of explanation achieved is, however, only

56

modest at fifty-three per cent, whereas a much higher proportion of attracted oil freight can be accounted for. But to achieve this, a motley collection of variables emerges; collectively, they represent a good sample of employment that is widely scattered, reflecting the widespread destinations to which oil is consigned for a very wide range of uses.

Transport equipment. The high degree of explanation achieved for the volume of originating freight is based on variables that in the main are plausible – metal manufacture (X_5), vehicles (X_8), metals not elsewhere specified (X_9) and other manufactures (X_{16}). Two seem distinctly odd and cannot be taken too seriously, namely food, drink and tobacco (X_3) and gas, electricity and water (X_{18}). A similar comment applies to the attraction of freight, though in this case the X_3 variable is the only odd one and the inclusion of transport and communications (X_{19}) obviously makes sense.

Food. The only odd feature is the inclusion of transport and communications (X_{19}) in the estimating equation for traffic generation. As this commodity class includes live animals, the attraction exerted by agriculture, forestry and fishing (X_1) is quite acceptable.

Chemicals. Both chemicals (X_4) and metal manufacture (X_5) are reasonable variables to explain both the generation and the attraction of this freight, while food, drink and tobacco (X_3) quite reasonably appears in the equation for attraction. However, only a moderate level of explanation is achieved ($R^2 = 0.65$ and 0.75).

Building materials. With a reasonably high level of explanation for both the generation and the attraction of building materials, the explanatory variables are eminently sensible. Agriculture, forestry and fishing (X_1) is related to the supply of timber, while mining and quarrying (X_2) is associated with the production of cement, clay products including bricks, sand and gravel. Since building supplies are often drawn from a very small radius, the appearance of construction (X_{17}) in both equations is not surprising.

Other manufactures. With the exception of textiles (X_{10}), the numerous variables included in both equations clearly represent a widely distributed set of activities that can reasonably be associated with an essentially heterogeneous commodity class.

Other crude materials. This commodity class includes animal feeding-stuffs, fertilisers, timber, textile fibres and waste, and non-ferrous ores. Thus the connection with agriculture, forestry and fishing (X_1) and textiles (X_{10}) is to be expected. The association with food, drink and tobacco (X_3) in both equations, and with shipbuilding and marine engineering (X_7) for the attraction of freight, is less obvious.

Miscellaneous. As the rag-bag category, miscellaneous traffic includes furniture removals, mixed loads, empty containers, laundry, and railway smalls and parcels traffic. Consequently, both origins and destinations are widely dispersed and there is much cross-traffic. So it will not stretch the imagination to find a rationalisation for almost any set of employment variables thrown up by the regression equations.

57

Discussion of results

Disaggregation of the employment variable does provide a useful basis on which to estimate the volume of freight generated in and attracted to zones, for most commodities. By far the poorest performance is for iron ore and limestone, and the attraction of scrap. For all other commodities, over half the variance is explained by employment at the S.I.C. Order level and, in the great majority of cases over seventy per cent is thus explained. While these results are encouraging, they are not yet good enough to permit confident forecasting of freight volumes given estimates of the employed population. The question arises whether the estimating equations can be improved.

There are three ways in which such improvement might be achieved:

(1) greater disaggregation of the freight and employment variables,
(2) more carefully designed geographical units for recording the data,
(3) the introduction of additional or altered variables.

In the present context, the first two approaches must be ruled out, though it should be noted that employment can be sub-divided to the Minimum List level of 152 industry classes. It is on the third issue that we shall concentrate.

After the work reported in this chapter was substantially complete, we received the results of work done at the Department of the Environment (1971) on modelling road freight flows, using 1962 data for the same seventy-eight zones that we have employed. In this work, conducted by Ruth Heyman, an attempt was made to improve the performance of the estimating equations by collapsing some of the twenty-four S.I.C. Orders into composite industries. This was done where inspection of the inter-correlation matrix showed a high level of geographical association between two or more Orders. Regression equations based on the reduced number of employment classes did not perform so well as when the twenty-four Orders were used. It was also found by Heyman that the introduction of port variables yielded no improvement in the estimating equations. Another abortive exercise was to incorporate estimates of spatial variation in productivity, but the reason for lack of success in this enterprise is probably the inadequacy of the data, which are also only available at the standard region level.

Given that the Department of the Environment found no basis for improving the performance of the equations on the lines indicated in the previous paragraph, there was little point in our pursuing similar enquiries. On the other hand, evidence was found by the Department of the Environment for consistent over-prediction and under-prediction of freight volumes for some zones. Enquiry showed that these errors lay in some of the seventy-eight zones which were obtained from the original 107 zones used in the 1962 survey be division. Division was in some cases necessary to make the seventy-eight zones compatible with the geographical areas for which rail freight were available. Where a zone was divided, the freight flows for the original zone were allocated to the two new zones in proportion to their respective populations, except in London (see p. 32). In this way

58

basic flaw in the data for twelve out of the seventy-eight zones was brought to light. Some improvement in the results of the step-wise regressions could be expected on a re-calculation using a reduced number of zones, though this exercise was not in fact carried out by the Department of the Environment. For the present purpose, let us note that the errors due to sub-division of zones will be compensating errors for those pairs of zones which were derived from the same original zone in the 107-fold division of the country. Thus, in examining the residuals there will not be a systematic bias at the national scale, only local bias between pairs of areas.

Regression residuals, total road transport

In view of the kinds of problem discussed above, an exhaustive discussion of the regression residuals does not appear warranted. It must also be remembered that in the commodity matrices there are numerous empty cells and cells with only small values. Therefore, to obtain residuals for each commodity for each zone would involve an unreasonably large proportion of cases in which the residuals could not be treated as even approximately reliable. Consequently, in this section we return to the aggregate level of analysis, as employed in table 5.1. In that table, the most successful estimating equations are for road transport alone; population is marginally more successful than employment as a predictor of freight volumes. Therefore residuals have been obtained from the estimating equations for road transport regressed on total resident population.

Reference to chapter 2 suggests that areas which are remote from the main centres of population may engage in activities requiring a low freight tonnage per head, as one means of avoiding some of the disadvantages of a peripheral location. The converse proposition might also be expected to hold. On this basis, we would expect the estimating equations of table 5.1 systematically to overestimate the tonnage in peripheral areas and to underestimate it in central ones. Chapter 2 develops the idea of centrality in terms of potential accessibility, measured in·population- or employment-miles. As normally calculated, the figure for the summation of population-miles for each zone includes the population of the zone itself (divided by the estimate of intra-zonal distance, in this case the estimated mean haul). As the total populations of the zones have been used as the independent variable in the estimating equations from which the residuals have been derived, an element of spurious correlation is introduced. In examining total freight tonnages, this defect cannot be removed by eliminating the intra-zonal populations from the calculations, since this would destroy the basis of the regressions (but compare p. 96 regarding mean haul).

The residuals have been calculated as a percentage of the actual volume of freight. A positive residual indicates that the actual volume of traffic exceeds the 'predicted' amount, whereas a negative residual means that the real volume is less than expected. The residuals for both O_i and D_j tonnages were regressed on population-miles, estimated with a distance exponent of -1.0.

Freight flows and the British economy

$$O_i \quad Y = 68.9671 - 0.1710 \, X \quad R^2 = 0.08$$
$$F = 6.87$$

$$D_j \quad Y = 43.3786 - 0.1169 \, X \quad R^2 = 0.08$$
$$F = 6.89$$

where Y = residual tonnage as a percentage ($\pm$) of actual freight volume and X = population-miles in thousands (including intra-zonal population).

In other experiments, particularly those discussed in chapter 6, regressions were improved by a logarithmic transformation of the independent variable. The values for potential accessibility as measured by population-miles are not normally distributed and the logarithmic transformation may be expected to yield a better estimating equation than is obtained using natural numbers. So a further exercise was undertaken in which the residuals were regressed on the logarithm of population-miles (including intra-zonal population and using a distance exponent of -1.0). For both O_i and D_j traffic, the values of R at -0.29 and of R^2 at 0.08 are identical to those previously obtained. Clearly there is no improvement in the performance of the correlations, which are not acceptable at the ninety-nine per cent level. But furthermore, the fact that there is no change in the level of association reinforces the conclusion that there is no relationship between potential accessibility and the tonnage of freight per person.

It should be remembered that nearly eighty per cent of the variance in road freight tonnage generated by and attracted to zones is explained by the resident population. The fact, therefore, that the residuals are completely unrelated to potential accessibility proves beyond a doubt that there is no relationship between freight volume and location relative to the main centres of population and activity. In short, the peripheral zones do not seek to avoid some of the disadvantages of their location by concentrating on activities with a relatively low tonnage per person.

Though the formal examination of the residuals is thus abortive, their spatial distribution remains of interest, as shown in figure 5.1. The most striking feature of this map is the overestimation of freight tonnage in London and adjacent areas to the north and south in the densely populated, heavily urbanised London region. It is also noteworthy how the high negative residuals outside this region are found mainly in urban areas such as Lancaster (zone 22) and Carlisle (zone 19); the less extreme negative values are concentrated in two areas, one straddling the northern Pennines and the other the eastern end of the Southern Uplands – in both cases predominantly rural areas. Moderate positive residuals – implying that the actual tonnage exceeds the estimated amount – are confined mainly to rural areas such as the Welsh Marches, and quasi-rural areas like southern Cheshire. Figure 5.1 makes clear that the relative error is not normally distributed. Positive residuals are much more numerous but much smaller in magnitude than are the negative residuals. This indicates that the estimating procedure errs most seriously by overestimating the volume of freight generated for a limited number of zones.

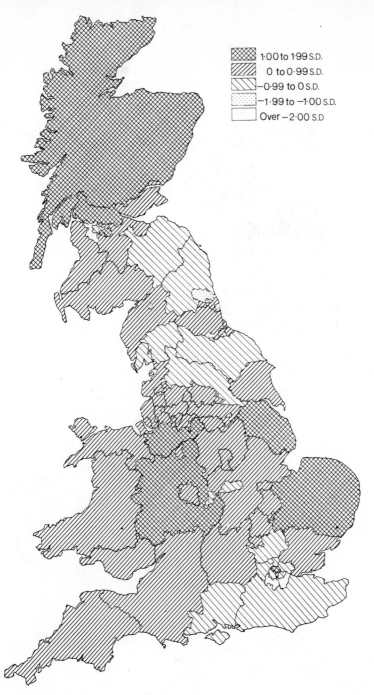

Fig. 5.1. Originating road freight: residuals from the regression of aggregate tonnage on resident population

61

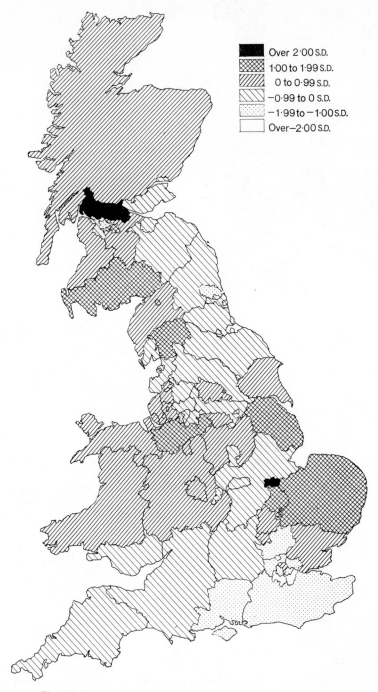

Fig. 5.2. Originating road freight: aggregate tonnage per resident

A slightly different way of approaching the matter is to obtain the tonnage of goods per resident; this has been done for road O_i traffic, as shown in figure 5.2. We should not expect the pattern to be identical to that shown in figure 5.1 and indeed it will be noted that there are substantial differences. The London area again stands out but now as part of a much larger swathe of territory running from Lands End to the Wash and thence south to Kent, in all of which the tonnage of originated road traffic *per caput* is low. A similarly extensive strip of country with low tonnages *per caput* runs from Manchester northwards to Dundee. As in Figure 5.1, there is no real suggestion that peripheral areas behave differently from other areas and there is less suggestion that urban areas are differentiated from all others.

In terms of the aggregate behaviour of road freight, there is thus clear evidence that the notion of centrality in the space-economy appears to be largely irrelevant as a factor affecting the volumes generated and attracted.

The spatial symmetry of tonnages originating and terminating, road freight only

Although for Great Britain as a whole the volume of originating freight is by definition equal to the volume that terminates, the same identity does not necessarily apply to each sub-area. If there are large differences in the volumes of O_i and D_j freight at the zonal level, two consequences may follow. First, in viewing the space-economy and the behaviour of freight traffic, it could make a substantial difference whether O_i or D_j freight patterns were considered. Secondly, there may be substantial regions of the country in which O_i traffic exceeds D_j, and vice versa; this would imply a net transfer of freight out of some areas and into others.

To test the symmetry of freight volumes, two exercises were undertaken for road freight only. For each of the seventy-eight zones the total O_i and D_j freight is known. This of course includes the intra-zonal freight, for which the tonnage originating equals the volume terminating. Interest therefore focusses on the inter-zonal freight which amounts to about forty-four per cent of the total road tonnage. A simple linear regression of tonnage O_i on tonnage D_j yields a value for R^2 of 0.94; in other words, the two patterns are almost identical. Consequently, we can conclude that for road freight there is no systematic net transfer of goods from some regions of the country into others.

The form of the estimating equation does, however, raise a most intriguing issue. If Y = the tonnage O_i and X the tonnage D_j (both in hundreds), then

$$Y = 3,742 + 0.952 \, X.$$

This indicates that where the freight volume is small, there is an excess of originating traffic, and vice versa. Now the volume of inter-zonal traffic is intimately related to the geographical extent of the zone as well as the aggregate amount of activity therein and there is thus no simple classification of zones that can be made. However, small volumes of inter-zonal traffic tend to be associated with zones of large extent, which tend to be rural in character. The converse is also true. There is, therefore, some hint that the

63

major urban centres consume a larger volume of road freight than they generate, which is consistent with the idea of agricultural products, building materials and raw materials being consumed or processed in the major industrial centres.

Conclusion

After the rather elaborate data manipulations described in this chapter, what conclusions can we draw? Perhaps the most striking point to emerge is that modelling rail freight volumes is much less successful than modelling road traffic; road plus rail freight is reasonably amenable to model building, largely because road traffic dominates most commodity groups. The second conclusion is that though the gross variables — population, employment and retail turnover — provide a good explanation of total freight volumes, they do not perform at all well for individual commodity groups. Thirdly, when attempts are made to improve the estimating equations for individual commodities by disaggregating employment with the twenty-four S.I.C. Orders, the performance is very good for some freights (mainly the more general classes), but poor for the mineral traffics. Given the data with which we have worked, it is not clear that working from the commodity classes and using the 24 industrial classes will yield a better estimation of total freight volumes than can be obtained by estimating directly from the more aggregate data. Indeed, the Department of the Environment (1971) found that if the aggregate road traffic O_i and D_j is regressed on employment in the 24 S.I.C. industries, ninety-three per cent of the variance can be explained.

Should we conclude that attempts to estimate freight volumes originating from and terminating in zones is only worthwhile at the most aggregate level? The answer would seem to be an unequivocal negative. In the first place, there would seem to be a strong case for treating road and rail freight separately. For the latter, it may be necessary to develop models specifically related to the main user industries, such as coal-mining and steelmaking, based on points of production and consumption. For road transport, on the other hand, a staged approach based on the techniques explored in this chapter is probably appropriate, though it has not been possible to proceed further along these lines in the present study. The approach we envisage is first to estimate the freight tonnage (O_i and D_j) for each zone at the aggregate level. Though on average this will give a fair estimate of the totals, individual zones will diverge rather seriously from the estimate so obtained. Therefore a second step would be to build up an estimate of the total volume for each zone (again O_i and D_j) on the basis of summing the estimates for individual commodities. This second set of totals would indicate for which zones the initial estimates were too high or too low. With admittedly heuristic adjustment, improved estimates should be obtained which would include a commodity breakdown.

Now the commodity composition of the freight is significant for two main reasons. The first is modal split. If road and rail are estimated separately this is not important. Secondly, within the road sector the commodity composition is important because it affects the number and kind of vehicles needed to move

64

a given volume of freight. This is a second-order question on which we need not dwell.

Although the results reported in this chapter do not open the way immediately to reliable forecasting of freight volumes in zones, they are sufficiently encouraging to warrant two final remarks. In the first place, it is quite clear that the level and nature of disaggregation of data, both by sector and by geographical zone, leave much to be desired. Intrinsically it is quite possible that much better forecasts could be made if more suitable data were available, though this would undoubtedly mean a much larger volume of material to handle and process. Finally, though population and employment totals are not themselves very amenable to forecasts, it is at least possible to make assumptions about the future spatial allocation of these strategic variables and then to engage in simulation exercises to examine the possible impact on freight volumes generated by and attracted to zones. We might envisage a number of feasible dispositions to accommodate the growing population of the nation and obtain some orders of magnitude for the freight volumes in zones that would result. The essence of the matter is that, given the manpower and computing resources, it would be feasible to engage in a series of simulation exercises to establish the range in values for freight generation and attraction, given assumptions about the geography of population and employment.

It is next necessary to establish efficient means for allocating the resulting flows. This is the problem with which we deal in the next chapter.

Chapter 6

THE DISTRIBUTION OF TRAFFIC

In chapter 2 we reviewed possible approaches to the modelling of freight flows and concluded that linear programming and the gravity model are the two most suitable frameworks within which to work, given the data that are available. Both of these techniques are applicable to a situation in which there is a finite set of origins and destinations, represented by the seventy-eight zones, for which information is available on total supplies and total demands. Though there are problems in forecasting the future supplies and demands (chapter 5), information is available for the position in 1964. In this chapter, therefore, we report the progress made in developing spatial allocation models, given information about the tonnage of goods originating and terminating in zones. The results are presented approximately in the sequence in which they were obtained, following a strategy similar to that employed in chapter 5, i.e., in the first instance work was done at a highly aggregated level and then subsequently at varying degrees of disaggregation.

The gravity model

In chapter 2, a brief derivation of the gravity model was given, yielding the following expression

$$T_{ij} = A_i . B_j . O_i . D_j . d_{ij}^{-\beta},$$ 6.1

where $A_i = [\sum_j B_j . D_j . d_{ij}^{-\beta}]^{-1}$ and $B_j = [\sum A_i . O_i . d_{ij}^{-\beta}]^{-1}$.

This is known as the production and attraction constrained version, which has two notable characteristics. It is primarily suited for modelling an entire national system, which is treated as being closed. It is also computationally very demanding. Therefore, for some initial experiments with only a sample of the origin zones, somewhat simpler versions were employed, in the form

$$T_{ij} = k . O_i . D_j . d_{ij}^{-\beta}$$ 6.2

k and β being parameters to be estimated. The other version allows the attracting force of the destination to assume an exponent,

$$T_{ij} = k . O_i . D_j^{\alpha} . d_{ij}^{-\beta}$$ 6.3

The parameters of equations 6.2 and 6.3 can be estimated by least squares if the expressions are made linear in logarithms. Thus

66

$$\log\left[\frac{T_{ij}}{O_i . D_j}\right] = a - b \log d_{ij}, \qquad\qquad 6.4$$

and

$$\log\left[\frac{T_{ij}}{O_i}\right] = a + b_1 \log D_j - b_2 \log d_{ij} \qquad\qquad 6.5$$

Equation 6.4 is the logarithmic equivalent of equation 6.2, while 6.5 similarly represents 6.3. In equation 6.4, antilog a gives k and b gives β: in equation 6.5, antilog a also gives k, b_1 gives α and b_2 gives β.

In modelling freight distribution by means of a gravity model, whatever its form, we came to the conclusion that there was little to be gained from applying the technique to rail traffic. The reasons for this conclusion will have become apparent in the preceding chapter. Consequently, all the work described in the present section relates to road freight only.

In the initial experiment, equations 6.4 and 6.5 were used. Because of a size limitation in the regression program available and the consequently extravagant use of computer time which would otherwise be necessary, this pilot project was confined to the first 20 rows of the matrix of total road traffic, taking each of the origin zones in turn. Using individual origins in this way, the O_i term is eliminated from the above equations. The column sums of the total road

Table 6.1. *The parameters of* $T_{ij} = k . D_j . d_{ij}^{-\beta}$ *for the first 20 rows of the total road tonnage matrix*

Origin zone	Centroid	log k	β	R^2
1	Morpeth	6.0978	−2.0	0.67
2	Tynemouth	5.2285	−1.6	0.64
3	Gateshead	5.5611	−1.8	0.42
4	Newcastle	5.1761	−1.7	0.47
5	Sunderland	7.2023	−2.7	0.60
6	Durham	6.4466	−2.2	0.51
7	Darlington	7.0092	−2.9	0.69
8	Stockton	5.5500	−1.8	0.47
9	Middlesbrough	7.2725	−2.8	0.63
10	Northallerton	6.9878	−2.7	0.38
11	Selby	2.1026	−2.7	0.38
12	York	8.4608	−3.5	0.74
13	Leeds	6.5225	−2.0	0.47
14	Barnsley	6.6426	−2.3	0.42
15	Doncaster	5.7115	−2.1	0.41
16	Sheffield	5.1324	−1.6	0.30
17	Hull	5.9426	−1.8	0.19
18	Workington	7.5855	−2.8	0.45
19	Carlisle	7.8388	−3.1	0.69
20	Kendal	7.7882	−3.0	0.53
Mean values			−2.3	0.50

Table 6.2. *The parameters of the 'Portbury' model $T_{ij} = k \cdot D_j^{\alpha} \cdot d_{ij}^{-\beta}$ for the first 20 rows of the total road tonnage matrix*

Origin zone	Centroid	log k	α	β	R^2
1	Morpeth	4.4832	0.4	−2.0	0.63
2	Tynemouth	2.8250	1.0	−1.9	0.80
3	Gateshead	−1.9506	1.6	−2.0	0.66
4	Newcastle	−1.8277	1.5	−1.9	0.68
5	Sunderland	−1.5912	1.4	−1.9	0.67
6	Durham	1.9819	1.0	−2.5	0.58
7	Darlington	0.5832	0.9	−1.9	0.62
8	Stockton	1.0439	1.0	−1.9	0.54
9	Middlesbrough	0.7345	0.9	−1.7	0.57
10	Northallerton	2.3300	1.1	−3.1	0.44
11	Selby	3.1929	1.0	−3.1	0.60
12	York	1.0155	1.0	−2.2	0.60
13	Leeds	1.9405	0.9	−1.9	0.61
14	Barnsley	1.3442	1.0	−2.2	0.52
15	Doncaster	2.1255	0.1	−0.6	0.06
16	Sheffield	0.9069	1.1	−1.3	0.40
17	Hull	6.1334	−0.9	−1.6	0.15
18	Workington	4.5326	0.8	−3.1	0.50
19	Carlisle	1.8573	0.7	−1.9	0.49
20	Kendal	3.3498	0.4	−2.0	0.25
Mean values			0.8	−2.0	0.50

tonnage matrix provide the D_j values and the elements of the matrix provide the values T_{ij}.

Tables 6.1 and 6.2 show the results of calibrating the two gravity models for the first twenty rows of the matrix. In both cases, there is a considerable range in the β values; in table 6.1, the extremes are −1.6 and −3.5, with a mean of −2.3; in table 6.2, the equivalent values are −0.6, −3.1 and −2.0 respectively. The levels of R^2 are comparable, with the mean value at 0.50 in both cases.

These initial experiments were sufficiently encouraging to warrant further exploration. Furthermore, the parameters displayed in tables 6.1 and 6.2 gave guidance in calibrating the full matrix model. Calibration is an iterative procedure whereby initial assumptions are made regarding the value of the parameters, based upon either a guess or the running of a linear version of the model using least squares, and successive trials are run until the parameters converge upon a stable set of values. Tests can include the correlation of actual and estimated T_{ij}s and trip length frequency distributions and the matching of actual and estimated mean lengths of haul. The process is essentially heuristic in character, and lacks a formal test that the 'best' fit has been obtained.

Equation 6.1 is computationally very demanding. As there appeared to be little to choose between 6.2 and 6.3 in terms of performance, and as the Mathematical Advisory Unit at the Ministry of Transport already had an

Table 6.3. Results from the production constrained 'Portbury' model, $T_{ij} = A_i \cdot O_i \cdot D_j^{\alpha} \cdot d_{ij}^{-\beta}$, calibrated with the 78 × 78 total road tonnage matrix, 1962

| Estimates with parameters | | Mean length of haul (miles) | Ton-miles (thousand million) | Trip length frequency distribution by distance categories (tons in miles categories) | | | | | | | | | | | R^2 actual on estimated T_{ij} | R^2 actual on estimated D_j |
α	β			0–25	26–50	51–75	76–100	101–25	126–50	151–75	176–200	201–25	226–50	Over 250		
1.0	−2.0	39	44.6	7,056,150	1,968,316	757,958	525,243	332,275	291,515	165,206	116,247	115,969	47,167	188,978	0.73	0.73
1.0	−2.5	24	27.4	8,000,826	1,995,563	609,750	327,291	190,032	165,666	79,942	51,140	47,574	18,552	78,845	0.76	0.75
1.0	−3.0	17	20.2	8,571,872	1,992,680	497,144	193,683	103,229	91,432	36,955	21,756	18,094	6,883	31,867	0.78	0.77
1.25	−2.0	39	44.8	6,991,534	2,011,899	776,866	530,909	336,001	287,091	164,370	112,403	117,625	46,944	189,433	0.74	0.76
1.25	−2.5	18	20.3	8,538,593	2,014,194	512,433	193,173	106,653	87,082	36,945	20,246	18,391	6,969	30,895	0.79	0.79
1.5	−2.0	40	47.5	6,893,209	2,051,974	802,156	543,157	344,276	286,189	166,269	111,106	122,342	48,576	196,049	0.74	0.77
1.5	−2.5	24	28.1	7,872,519	2,062,192	654,845	336,032	202,119	158,236	81,540	47,433	50,645	19,732	79,884	0.78	0.78
1.5	−3.0	18	20.6	8,486,349	2,035,353	533,207	196,593	113,416	84,605	38,247	19,558	19,448	7,609	31,227	0.80	0.80
Actual values		32	37.5	6,982,992	2,403,722	866,838	518,737	308,512	164,219	99,863	75,895	74,856	19,553	52,892		

Note: owing to rounding and the fact that the model is not fully constrained, the rows do not have an identical sum.

operational programme for 6.3, it was decided to use this version of the gravity model. The Mathematical Advisory Unit very kindly ran this programme with the 78 x 78 total road flow data for 1962. Equation 6.3 is known as the Portbury model, because it was used by the Ministry in examining the case for a major port near Avonmouth, Bristol (Ministry of Transport, 1966a). As already noted, it is a production constrained model in which the attraction term assumes an exponent. The constraint is

$$\sum_j T_{ij} = O_i,$$

which can be satisfied by replacing k in equation 6.3 by a set of factors, A_i

$$T_{ij} = A_i \cdot O_i \cdot D_j^{\alpha} \cdot d_{ij}^{-\beta}, \qquad\qquad 6.6$$

such that

$$A_i = [\sum_j D_j^{\alpha} \cdot d_{ij}^{-\beta}]^{-1}$$

The model is calibrated and tested by searching an array of combinations of α and β values, for which the A_i, and hence the D_j and T_{ij}, values are computed. The estimated D_js for each combination of α and β can be compared with the actual values. Similarly, the estimated T_{ij}s can be compared with reality in terms of the mean haul and the distribution of tonnages over distance bands. Results, using equation 6.6, are set out in table 6.3 for a range of values of both α and β. It will be noted that the values of R^2 obtained are uniformly good, the lowest value being 0.73 and the highest 0.80. However, this result is somewhat disconcerting because the values of R^2 do not vary sufficiently for a confident decision to be taken that one particular combination of parameters is undoubtedly the best. The other feature to note is that even though a good explanation of the flows (T_{ij}) and attractions (D_j) is obtained, both the mean haul and the summation of ton-miles differ substantially from the actual values. The difference is not a systematic over- or underestimate, but is primarily related to the selected value for the distance exponent, β: the larger the β coefficient, the shorter the estimated mean haul and the smaller the ton-mileage.

Overall, it appears that the best result is given by an α of 1.25 or 1.5 and a β of -2.5. This β is near the middle of the range obtained in table 6.2, but the α differs markedly. On the other hand, a β of -2.5 lies neatly in the middle of the range of values from -2.0 to -3.0 that Kirby (1969) obtained in collating distance friction factors from studies of inter-urban traffic movement in Great Britain.

Spatial variation in distance exponents

Tables 6.1 and 6.2 indicate clearly that the distance exponents, β, vary considerably from one zone to another. It is, therefore, advisable to explore systematically for all seventy-eight zones just how stable the values for β are. This exercise was undertaken for each row of the matrix, i.e., for all origins.

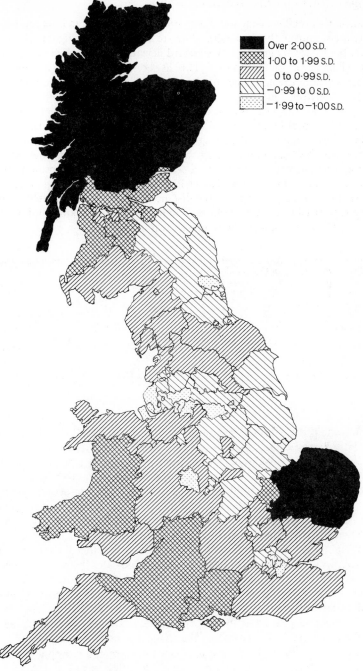

Over 2·00 S.D.

1·00 to 1·99 S.D.

0 to 0·99 S.D.

−0·99 to 0 S.D.

−1·99 to −1·00 S.D.

Fig. 6.1. Originating road traffic: gravity model distance exponents (β)

71

Freight flows and the British economy

Equation 6.4, the linear version of equation 6.2, was used and the resulting β values are set out in the Appendix: they are also mapped in figure 6.1. In this figure, the values are grouped by standard deviations about their mean value of −2.4, which is very similar to −2.5. This, as shown in table 6.3, is a value that performs well in fitting a single model to all origins and destinations simultaneously. Though this mean conforms very nicely with Kirby's (1969) findings, the range about the mean is considerable − from a low value of −1.3 in zones 48 and 49 (central London) to −4.8 in zone 68 (northern Scotland).

Spatial variation in the distance exponents suggests a certain urban/rural duality, with low values in the urban areas and high ones in the rural parts of the country. There also seems to be a general accessibility dimension in the variation, with peripheral areas having high values and central regions low ones. To examine this pattern further, the distance exponents were treated as the dependent variable and were regressed on population-miles and employment-miles $d_{ij}^{-1.0}$, see p. 7), with the results shown in table 6.4.

Table 6.4. *Road freight: the relationship between gravity-model distance exponents (β) and potential accessibility (using $d_{ij}^{-1.0}$)*

Independent variable	a	b	R^2	F
Population-miles (000)	2.9145	−0.0008	0.29	31.33
Log_{10} of population-miles	8.3820	−2.1654	0.48	70.89
Employment-miles (000)	2.8800	−0.0144	0.25	25.54
Log_{10} of employment-miles	4.8920	−1.7190	0.39	48.82

Note: the dependent variable (β) is negative.

Not only is the expected negative relationship manifest, but the level of association is respectably high. It will be seen that population-miles perform somewhat better than employment-miles as a measure of general accessibility, though the difference is not dramatic. Perhaps more useful to note is that the logarithmic transformation of the independent variable improves the performance of the equations quite considerably. Thus, the logarithm of population-miles 'explains' forty-eight per cent of the spatial variation in the distance exponent. It is abundantly clear, therefore, that the traffic patterns of peripheral regions do differ from the central areas, and that the interpretation is as follows. The distance exponent for a zone indicates the rate at which inter-action would decline with distance, given the spatial distribution of opportunities for transactions, if all other conditions were equal. A large exponent, therefore, implies that a zone has a pattern of freight movement based on a relatively large element of local self-sufficiency; since high exponents are associated with peripheral locations, it appears that the remoter zones attempt to avoid the penalties of long hauls to the central zones. Conversely, the central areas, having low exponents, distribute their traffic with less regard to distance.

It is clear that factors other than general accessibility are also important in

Fig. 6.2. Originating road traffic: residuals from the regression of β coefficients on population-miles

73

explaining the spatial variation in the β values. In particular, one would expect the commodity composition to be an important variable — and indeed, some of the coalfield zones and the East Midlands iron ore fields do have low coefficients. Unfortunately, the matrices for individual commodities have many empty cells, so that it really is not practicable to fit gravity models for each commodity for each zone (compare p. 59). Therefore, because the analysis of the spatial variation of the β values cannot be pressed further with the data available to us, we content ourselves with showing the spatial variation of the residual values of β obtained from the estimating equation in table 6.4 relating the distance exponents to the logarithm of population-miles (figure 6.2). Perhaps the striking thing about this map is the large area in southern and central England, all of Wales and most of central Scotland, in which the actual values for β are greater than is predicted from the estimating equation. Zones with a distance friction less than would be expected lie mainly in northern England and southern Scotland.

Comparing these residuals with the data in table 3.6, two things are apparent. Of the thirty-five zones with ports engaging in oversea trade, twenty-two have actual distance exponents that exceed the predicted values. Thus, although port zones occur in all six of the classes identified in figure 6.2, there is a preponderance of them in the zones that experience a distance friction greater than expected. Furthermore, all the port zones with really substantial foreign trade at the top of table 3.6 and as far down as zone 65 (Cardiff) have actual distance exponents greater than the expected values. Zone 31 (Manchester) is the first one reached to have the actual distance friction less than expected, and Manchester is an inland port zone in the centre of a major conurbation. This evidence clearly confirms the conclusion reached in chapter 3 that the incidence of foreign trade on inland freight movement can be ignored and in any case trade effects tend to operate in the opposite direction to that which might be expected.

Two conclusions may be drawn from this examination of the β values of the gravity model. The first is that the considerable spatial variations in the distance exponents can in substantial part be explained by reference to the concept of potential accessibility, measured by population-miles. Furthermore, the nature of the relationship is such that it confirms the hypothesis advanced in chapter 2, that peripheral areas will try to avoid the penalty of long hauls to the major centres of activity by a greater degree of local self-sufficiency. The second and related conclusion is that given this spatial variation in exponents, a single model applied to the whole country may result in some serious over- or under-estimating of traffic. That these errors are not large in aggregate terms is shown by the high level of R^2 achieved; but for zones with small freight volumes, the errors are clearly going to be proportionately very large in some cases (but see chapter 10).

Commodity group variations in distance exponents

The next step was to see whether the gravity model can usefully be applied to the commodity groups. For this purpose, equation 6.4 was again used as the operational form of equation 6.2. The results are set out in table 6.5. As can be seen, the level of explanation achieved never rises above sixty-two per cent (for food) and falls as low as twenty-four per cent for steel. The distance exponents range from a high value for building materials at −1.6 and a low at −0.8 for steel, scrap and chemicals. The ranking of the commodities is intuitively sensible and accords with evidence adduced by Chisholm (1971a) on the geographical range over which various classes of commodity are moved. There appears to be a slight tendency for more highly valued and processed goods, such as steel, transport vehicles and equipment and chemicals, to have low values as opposed to the higher values for bulky goods such as coal and coke and building materials, but it is not very marked.

Table 6.5. *Road freight: estimates of parameters of* $T_{ij} = k \cdot O_i \cdot D_j \cdot d_{ij}^{-\beta}$ *for individual commodities*

	Commodity	log k	β	R^2
1	Coal and coke	−8.887	−1.5	0.49
4	Scrap	−8.866	−0.8	0.28
5	Steel	−7.758	−0.8	0.24
6	Oil	−8.075	−1.2	0.37
7	Transport goods, etc.	−6.917	−1.0	0.41
8	Food	−8.734	−1.5	0.62
9	Chemicals	−8.953	−0.8	0.31
10	Building materials	−9.196	−1.6	0.60
11	Other manufactures	−9.304	−1.1	0.43
12	Other crude	−8.572	−1.2	0.45
13	Miscellaneous	−8.482	−1.4	0.51

The small range in the β values for commodities compared with the zonal variation for aggregate freight is probably due to the much higher level of final aggregation, in which the more extreme values tend to cancel out. It is much more difficult to explain why all the commodity exponents lie below the mean of the zonal β values. The explanation may lie in the fact that for each commodity the possible number of flows is 78^2, whereas in calibrating the gravity model for each zone only seventy-eight observations are used, but it is not immediately apparent why this should be the case.

Conclusions on the gravity model

The performance of the gravity model approach to the allocation of freight flows between pairs of areas clearly leaves much to be desired. Although a good fit can be obtained with the national system of aggregate freight flows, it is worrying that the model is comparatively insensitive to variations in the parameter values. It is equally disturbing that there is such a large variation in the zonal values for the distance exponents, though this fact does throw interesting light on one proposition that can be derived from location theory. Finally, the

Freight flows and the British economy

relatively low values for R^2 for the commodity classes indicate that the gravity model is not very suitable for examining flows disaggregated by type of goods.

A rather similar conclusion was reached by Frost (1969), using the same data that we have employed. He selected commodity class 11, other manufactures. Most of the cells of this commodity matrix are occupied and this hetero-geneous class of goods is distributed over considerable distances, unlike building materials (table 6.11). On the other hand, 150 million tons distributed over a zonal system of seventy-eight areas implies comparatively small T_{ij} values on average — and hence the errors of sampling referred to on p. 32. Bearing in mind these reservations, Frost obtained β values for all seventy-eight origins, using a model in the form

$$T_{ij} = k . P_j^{\alpha} . d_{ij}^{-\beta}, \qquad\qquad 6.7$$

where P_j is the population of the destination zone. His estimates of β varied from -1.2 to -5.2 and the amount of variance in the flows explained by the model never rose above fifty-six per cent.

The Department of the Environment (1971) has also experimented with gravity models, using equation 6.1 and road freight flows for the 78 x 78 matrix. This work was started after we had embarked on our own study and led to the same conclusion — that the gravity model is not very efficient and that there-fore attention should be directed toward linear programming as an alternative approach.

However, if 'the most probable distribution' of the gravity model is not an appropriate framework within which to work at the level of aggregation here employed, more encouraging results have been obtained by Black (1971) for the United States. He used a model similar to equation 6.6 but without raising D_j to a power (α) and fitted it to data for twenty-four commodity classes moving between the nine census regions. His data therefore were more aggregated spatially but more finely differentiated by commodity than ours and he ignored intra-regional movements. The variance explained by the model ranged from seventy-three per cent for petroleum and coal products to ninety-nine per cent for textile and leather products. The β values varied from -0.25 for non-electrical industrial machinery to -5.3 for stone, clay and glass products. Though the values ranged more widely, the general tendency for more highly processed goods to have lower exponents and bulkier, low value goods to have higher values, is in accord with the findings presented here. However, it is unwise to compare too closely the results of studies with such enormously different areal bases. Considering the fineness of Black's commodity classifications and the probable distinctness of his regional units, one wonders if a model based on economic theory would not have been more appropriate. So we now turn to the linear programming approach to see whether, with the British data, a better level of explanation can be achieved than with the gravity model.

Linear programming solutions

A previous and fairly successful application of linear programming (L.P.) to the prediction of total freight flows in Ireland (O'Sullivan, 1968, 1969) encouraged

76

an exploration of this avenue in the British context. It seemed that simple
Transportation Problem solutions might produce good estimates of future
traffic distribution. The Transportation Problem is formally written out in
chapter 2, but it is convenient to re-state it briefly here. Given a set of
quantities of a good available at origins (O_i) and a set of requirements at
destinations (D_j) and the costs of transport between origins and destinations
(d_{ij}), the problem is to find a set of non-negative flows between origins and
destinations (T_{ij}) which will fulfil requirements and respect availabilities while
simultaneously minimising the total cost of transport. Making the simplifying
assumption that $\sum_i O_i = \sum_j D_j$ we can write the primal, minimising problem as:

$$\text{Min } C = \sum_i \sum_j T_{ij} \cdot d_{ij}, \qquad\qquad 6.8$$

subject to:

$$\sum_i T_{ij} = D_j,$$

$$\sum_j T_{ij} = O_i,$$

$$T_{ij} \geqslant 0.$$

The maximising dual to this problem can be written:

$$\text{Max } V = \sum_j V_j \cdot D_j - \sum_i U_i \cdot O_i \qquad\qquad 6.9$$

Where U_i and V_j are the solution 'shadow prices' associated with production at
origins and requirements at destinations – reflecting relative locational advantages
with respect to transport costs. The shadow prices obtained in solving the problem
are thus indicative of geographical comparative cost advantages for production
and consumption. A general description of the Transportation Problem and its
solution is given in chapter 5 of Dorfman, Samuelson and Solow (1958). The
economic interpretation of the dual values is discussed fully by Stevens (1961).

The row and column totals of the eleven road and thirteen rail commodity
flow matrices were used as availabilities (O_i) and requirements (D_j) and road
distances were used as a surrogate for transport cost (d_{ij}). As a first experiment
in this field, transportation problems were solved for movements by road of
the eleven commodities between seventy-eight regions – including intrazonal
movements. Then, to get a more comprehensive picture for the more homogeneous commodities with large volumes carried by rail – including those
carried exclusively on the railways – solutions were found for total road and
rail flows between seventy-eight regions for the first six commodity classes.
Note that we have not used the L.P. approach to model the aggregate flow of
all goods and to this extent we have departed from the strategy previously
employed – of working from the more aggregate to the less aggregate level. The
reason is that conceptually the L.P. approach assumes rational behaviour by
economically motivated men dealing with a homogeneous commodity. Therefore, it is fundamentally inappropriate to apply the L.P. technique to the
heterogeneous aggregation of all goods.

Freight flows and the British economy

Road distances were used in the calculations as they were available and much effort would have had to be expended to extract rail distances and decide on the appropriate combinations of the two. The computations were carried out with a programme written by John Adams of University College, London, which uses the MODI. method of solution.

The seventy-eight zones vary widely in area and industrial characteristics, a fact which obviously distorts the picture presented by the data. To see if the level of explanation achieved improved if a more homogeneous sub-set of zones were used, solutions were found for road flows of eleven commodities between twenty-four major urban areas of England and Wales. This exercise was carried out by Michael Dunford (1969).

Actual flows and L.P. solutions were compared by correlation and regression. The regression of L.P. on actual flows suggests the manner and magnitude of the departure of one from the other. Perfect correspondence would result in an intercept value of zero and a slope of unity. As a further comparison, the number of occupied cells in the actual origin/destination matrices were counted to match against the m + n − 1 positive values that are obtained from L.P. solutions for m origins and n destinations. These were expressed as percentages of the total number of cells (m x n).

Table 6.6 shows the comparison between actual freight flows and the L.P. solutions, for road traffic only. Apart from coal and coke and miscellaneous,

Table 6.6. *Road freight: comparison of actual flows with linear programming solutions, individual commodities*

	Commodity	R^2	Intercept	Slope	% cells occupied in actual matrix
		Regression of L.P. on actual			
1	Coal and coke	0.39	541	0.54	18
4	Scrap	0.87	52	0.70	9
5	Steel	0.89	−87	1.41	37
6	Oil	0.69	−46	1.03	20
7	Transport goods, etc.	0.79	14	0.56	40
8	Food	0.82	402	0.62	40
9	Chemicals	0.51	−33	1.29	33
10	Building materials	0.81	149	1.04	41
11	Other manufactures	0.57	−94	1.15	61
12	Other crude	0.82	−61	1.18	35
13	Miscellaneous	0.20	295	0.28	41
m + n − 1 as a percentage of total cells					2.5

the correspondence between the real and estimated flows is good, values of R^2 ranging from 0.51 to 0.89. Again with the exception of coal and coke, the more nearly a commodity is homogeneous in character, the better is the L.P. solution, a result that one might reasonably expect. In the case of coal and coke, the

78

poor performance of the L.P. solution probably arises from the fact that much coal and coke is in fact moved by rail, so that road movements may be an erratic sample from the total flow for this commodity group.

One striking point about table 6.6 is that the general level of R^2 is substantially higher than is achieved for the same commodities allocated by means of the gravity model (table 6.5) – a mean of 0.67 compared with 0.43. However, it should be noted that the gravity model R^2 values are obtained from estimating equations in linear form, while the L.P. values are correlations of actual and expected flow matrices. It is also interesting to note that there is some evidence for an inverse relationship between the two results; the higher the L.P. level of explanation, the lower the gravity model one and vice versa. However, two commodity groups are notably out of line, namely food and building materials. Both are characterised by short hauls and widespread sources and destinations, so perhaps the high levels of explanation achieved by both models are not surprising.

No comparison of the results in table 6.6 with comparable studies by other workers is possible because, so far as we are aware, none have been carried out. The Department of the Environment (1971) solved the transportation problem for coal and coke only. As with our results, the outcome was very disappointing and they did not proceed to apply L.P. to the other commodity classes. On the basis of the trial with coal and coke, it was concluded by the Department that the gravity model performs better than L.P.: while this conclusion is valid for road transport in this one commodity class, it clearly is not generally true.

Turning to table 6.7, showing the L.P. solutions for the six most homogeneous commodities, road and rail traffic combined, it is apparent that a much better result for coal and coke is now obtained. For the other three commodities common to both tables 6.6 and 6.7, the values of R^2 are slightly lower with rail freight included than when road transport only is considered. On the other hand, iron ore and limestone, both of which move exclusively by rail, achieve a level of explanation comparable with the best results for road alone, at about ninety per cent of the variance.

A feature common to both table 6.6 and table 6.7 is the concentration of the L.P. traffic into a very small number of flows. In other words, the L.P. solution eliminates a very high proportion of the small cross flows. This suggests that if the transportation problem is solved for a sub-set of zones comprising the major urban areas, a much better fit can be obtained than when L.P. is used on all seventy-eight zones. This expectation is amply confirmed by table 6.8, showing the L.P. results for twenty-four zones representing major cities; the level of explanation achieved falls below fifty per cent in only one case – other manufactures – and in all other cases is in excess of ninety per cent. To obtain the results shown in this table, Dunford restricted his attention to road traffic and to the urban areas in England and Wales. Furthermore, he selected only those urban areas which are not included in zones of large geographical extent; for example, zone 60, with Wolverhampton as its centroid, was excluded because it is very extensive and cannot be regarded as a truly urban area. In the second place, by referring to the 1962 survey data, Dunford was in some cases able to disaggregate

79

Freight flows and the British economy

Table 6.7. *Road plus rail traffic: comparison of actual flows with linear programming solutions, selected commodities*

	Commodity	Regression of L.P. on actual R^2	Intercept	Slope	% cells occupied in actual matrix
1	Coal and coke	0.85	−858	1.14	26
2	Iron ore	0.91	4	0.99	1
3	Limestone	0.89	−14	1.04	2
4	Scrap	0.82	−38	1.24	18
5	Steel	0.88	−97	1.42	45
6	Oil	0.70	−87	1.16	21

For 1, 4, 5 and 6   m + n − 1 as % of total cells   2.5
For 2   m + n − 1 as % of total cells   0.5
For 3   m + n − 1 as % of total cells   0.2

Table 6.8. *Road freight: comparison of actual flows between 24 major cities with linear programming solutions, individual commodities*

	Commodity	Regression of L.P. on actual R^2	Intercept	Slope	% cells occupied in actual matrix
1	Coal and coke	0.99	−129	1.03	27
4	Scrap	0.98	−43	1.08	21
5	Steel	0.96	−141	1.21	53
6	Oil	0.98	−153	1.11	26
7	Transport goods, etc.	0.94	−21	1.23	58
8	Food	0.96	−525	1.18	58
9	Chemicals	0.91	−139	1.27	48
10	Building materials	0.96	−247	1.07	49
11	Other manufactures	0.49	−463	1.33	77
12	Other crude	0.97	−139	1.16	45
13	Miscellaneous	0.97	−254	1.17	56

m + n − 1 as a percentage of total cells   8.3

Centroids of zones (see text)

Birmingham	Hull	Norwich
Bristol	Leeds	Nottingham
Bromley	Leicester	Plymouth
Cardiff	Leyton	Sheffield
Coventry	Liverpool	Southall
Croydon	Manchester	Southampton
Harringay	Middlesbrough	Stoke-on-Trent
Holborn	Newcastle	Swansea

Source: Dunford, 1969.

The distribution of traffic

large zones to distinguish the major urban areas contained therein. In this way, Bristol is separated from the rest of zone 58 and Cardiff and Swansea distinguished within zone 65. Therefore the zone system used in table 6.8, though employing the same centroids as for the 78-zone system, is not identical.

There seems to be little doubt that L.P. is much superior to the gravity model formulation for describing commodity flows. The superiority is such that at least for the major urban areas there can be no question but that the L.P. approach is to be preferred. However, the solutions obtained using the 78 x 78 matrices give a much higher concentration of flows than is actually observed. A solution to this problem can be envisaged in the following manner.

Linear programming and the Poisson distribution

A major drawback to using L.P. for predicting at the scale at which we are working here is the fact that the m + n − 1 solution values for the most part fail to give anything like the dispersal of actual traffic. It is noteworthy that for the most dispersed, heterogeneous commodity class (miscellaneous) a gravity model provides a closer image of reality. If we compare tables 6.5 and 6.6 we get a hint of an inverse complementarity between the two methods − the better the one, the worse the other. An alternative method, perhaps lying near the point of convergence of the two approaches, has been suggested by Morrill (1967) in a personal travel context. If we know that errors of decision on the actors' part are represented in the pattern we are trying to replicate and explain, we could assume the errors to be distributed according to a Poisson function, and disperse flows about their optimal paths accordingly. This can be accomplished in the L.P. framework by finding the optimal routes and allocating sixty-eight per cent of flows to these; these optimal paths are then prohibited for a second solution which produces second-best routes, to which twenty-seven per cent of flows can be allocated; finally, optimal and second-best routes are prohibited and third-best solutions are found, to which the balance of flows is allocated. Leigh Herrington (1971) undertook an experiment along these lines with the flows by road between the twenty-four major cities in England and Wales of the commodity other manufactured goods. A comparison of the Poisson-modified L.P. solution with the actual flows showed no improvement in terms of correlation of the elements of the matrices or of trip length frequency distributions over the straight L.P. solution results. However, there was an increase in the number of cells occupied from ten per cent of the actual cell occupations for the L.P. solution to thirty-two per cent for the modified version.

Dual values

Dual values obtained from the linear programming solutions can be interpreted as shadow prices and, therefore, as a measure of competitive advantage or disadvantage. However, the interpretation is fraught with considerable difficulty, not only on the grounds of the conceptual significance that can be attached to

Shadow Prices
Quartiles

Lowest

Highest

Fig. 6.3. Road plus rail: shadow prices for coal and coke − origins

Shadow Prices
Quartiles

Lowest

Highest

Fig. 6.4. Road plus rail: shadow prices for coal and coke — destinations

83

Fig. 6.5. Road plus rail: shadow prices for iron ore — origins

84

Shadow Prices
Quartiles

Lowest

Highest

Fig. 6.6. Road plus rail: shadow prices for iron ore – destinations

85

Shadow prices
Quartiles

Lowest

Highest

Fig. 6.7. Road plus rail: shadow prices for limestone − origins

Fig. 6.8. Road plus rail: shadow prices for limestone – destinations

Shadow Prices
Quartiles

Lowest

Highest

Fig. 6.9. Road plus rail: shadow prices for scrap — origins

Fig. 6.10. Road plus rail: shadow prices for scrap — destinations

Shadow Prices
Quartiles

Lowest

Highest

Fig. 6.11. Road plus rail: shadow prices for steel − origins

Fig. 6.12. Road plus rail: shadow prices for steel – destinations

91

Shadow Prices
Quartiles

Lowest

Highest

Fig. 6.13. Road plus rail: shadow prices for oil – origins

92

Fig. 6.14. Road plus rail: shadow prices for oil — destinations

93

the dual values, but also because of the nature of the data we are using. The commodity classes are not homogeneous and the origin-destination zones vary considerably in size. Perhaps more important, though, is the fact that traffic flows represent the working of the transport system and not the complete paths of goods from point of output to point of consumption.

Examining the dual values for road movements alone is a dangerous enterprise, since for many commodities there is a significant movement by rail and a not altogether clear relationship between length of trip and mode used. On the other hand, when the dual values for total movements are considered at least some of the distorting effects are overcome. Thus attention is confined to the dual values for freight moving by road and rail combined. The results are shown in map form, with the zones allocated to groups representing quartile divisions of the scale from highest to lowest value.

Some of the distributions of values in figures 6.3 to 6.14 do make geographical sense and there is an encouraging spatial clustering of values for zones of similar locational and/or activity characteristics, despite disparities in zone size. For coal and coke (figures 6.3 and 6.4), the transport system operations reflected in the flows confuse the picture, but for both origins and destinations prices are high in areas remote from major coalfields, i.e. southern England, and decrease with closeness to the sources of coal. The values for iron ore (figures 6.5 and 6.6) at origins and destinations do generally increase with distance from the domestic sources around Northampton, but the movement of imported ores complicates matters. Sources of limestone (figures 6.7 and 6.8) are quite widespread in Great Britain and the shadow prices vary with general accessibility; in central areas origin prices are low, whereas in peripheral areas they are high. Scrap prices (figures 6.9 and 6.10) seem to be higher at origins with heavy demand, i.e. steel producing areas: supply areas in southern England with little steel production have low values. Prices for steel (figures 6.11 and 6.12) are low at production points and in consuming areas close to them and rise through central and southern consuming areas. The picture presented by the values for oil (figures 6.13 and 6.14) is the most heartening justification of this effort. For this fairly homogeneous commodity, moving from a few coastal importing and refining points scattered over the country, low prices at these origins and the increase in price with distance from them seems sensible. The same pattern holds for prices at destinations – the more remote the area from the major refinery locations, the higher the prices.

Comparison of gravity model and linear programming solutions

As indicated on p. 81, it is evident that the gravity model approach in general performs less well than the linear programming solutions in describing the actually observed flows. For purposes of prediction, therefore, it would seem wiser in the future to concentrate efforts upon improving the performance of the L.P. solutions, perhaps along the lines explored, unsuccessfully so far, of incorporating a probabilistic element into the L.P. technique. An alternative might be a two-stage approach, in which the flows between major cities could be modelled by linear programming. Around the substantial framework so established, a second stage might be to develop a version of the gravity model to account for aggregate flows to and from the remaining zones. If such a strategy, or one similar to it, were adopted, a first task would be to experiment with the system of major cities to discover how far down the hierarchy it is possible to proceed without serious loss of efficiency. An alternative strategy may be to model some commodities by L.P. and some by means of the gravity model.

Though the evidence suggests that the L.P. format is to be preferred to the gravity model, it must be remembered that this conclusion is specific to the geographical scale and level of commodity aggregation employed.

Other aspects of distance friction

In chapter 2 it was suggested that distance friction could be measured in more than one way. Of the measures suggested, the gravity model β coefficients have already been examined for road freight and, encouragingly, nearly half the spatial variation can be accounted for by potential accessibility defined in population-miles. In this section, therefore, we examine the other measure that was mentioned – mean haul. For this purpose, attention will be directed exclusively to road transport, since it has become abundantly clear that the spatial distribution of rail traffic is dominated by the locations of a few relatively big sources of supply and points of consumption. General modelling of freight flows by rail is clearly not likely to be very rewarding except in the linear programming format. Road transport, on the other hand, appears to be much more amenable to generalised analysis. For reasons given in chapter 4, the analysis can be conducted only at the aggregate level; disaggregation by commodity is not really feasible.

Spatial variations in mean haul

The mean haul for traffic for any zone is obtained by the following expression

$$\frac{\sum\limits_{j} T_{ij} \cdot d_{ij}}{\sum\limits_{j} T_{ij}},$$

where T is the flow in tons and d is the distance in miles, for all pairs of flows, including the intra-zonal movement (j = i). The results, which are listed in the Appendix, were correlated with population-miles ($d_{ij}^{-1.0}$), the outcome being shown in table 6.9.

Though all four regressions are acceptable at the ninety-nine per cent level, the semi-logarithmic regressions perform better than those computed in natural numbers exclusively. About one-quarter of the variation in mean haul can be attributed to potential accessibility as measured by population-miles. In all cases, the sign of the coefficient is negative, indicating that the larger the value for population-miles the shorter the mean haul. This accords with the *a priori*

Freight flows and the British economy

Table 6.9. *The relationship between mean haul in miles for total road traffic and potential accessibility (using $d_{ij}^{-1.0}$)*

Independent variable	a	b	R^2	F
Traffic generated (O_i)				
Population-miles (000)	36.6151	−0.0067	0.13	11.81
$\log_{10}$ of population-miles	88.2899	−20.3044	0.26	27.29
Traffic attracted (D_j)				
Population-miles (000)	34.9051	−0.0051	0.09	7.22
$\log_{10}$ of population-miles	84.0968	−19.0165	0.26	26.49

Note: For O_i traffic, the logarithm of employment-miles yields an R^2 of 0.17, F = 15.71.

postulate set out in chapter 2. On the other hand, it may be thought that these results conflict with the evidence of table 6.4 in that a long mean haul is clearly associated with a large β value; the correlation of these two variables for total road originating traffic (O_i) is, in fact, moderately high, with an R^2 of 0.31, but negative in sign. The interpretation to be put on this finding is as follows. Zones which have a low value for population-miles tend to have a high distance friction, represented by a large distance exponent. To this extent they avoid the penalties of long hauls. But the adjustment is incomplete, because the local economy is unable to supply all needs and, therefore, some transactions with distant zones must take place. Consequently, the average length of haul for zones with a low population-miles figure tends to be higher than elsewhere. Thus we have evidence that the freight transactions of zones are adjusted at least in part to location within the space-economy, and that the adjustment occurs simultaneously both in the degree of attempted local self-sufficiency and in the mean haul for freight moving. However, whereas about half the variation in β coefficients can be so explained, only one-quarter of the mean haul variation is accounted for.

As calculated by the formula on p. 95, the mean length of haul for a zone includes the intra-zonal traffic, which amounts to fifty-six per cent of the aggregate tonnage and twenty-seven per cent of the ton-miles. Estimation of the mean haul for this intra-zonal traffic is extremely crude, as indicated on p. 34. This suggests that there is a case for examining the spatial distribution of the inter-zonal mean haul, the intra-zonal traffic being excluded. In this way we can eliminate the effects of inaccurate estimation of the length of haul for the within-zone traffic. By so doing, we may expect to improve the relationship between mean haul and potential accessibility. By eliminating the intra-zonal traffic, we may also dispose of an element of spurious correlation. In general, the zones of lesser geographical extent are heavily urbanised, either as parts of conurbations or as free-standing cities. Conversely, the larger zones are mainly rural. Therefore the size of zone and hence the estimate of intra-zonal mean haul may be expected to vary inversely with potential accessibility; the greater the value for population-miles, the shorter the intra-zonal haul, and vice versa.

Table 6.10 shows the relationship between the mean haul for inter-zonal road traffic, i.e., all freight movements except the intra-zonal consignments. If this table is compared with table 6.9, it is striking that the level of explanation achieved is some fifty per cent greater, the value of R^2 being 0.40 instead of 0.26. Clearly, for that traffic which does cross zonal boundaries the question of location within Britain does have a substantial impact upon the mean haul, and in the manner that theory leads us to expect.

Table 6.10. *The relationship between mean haul in miles for total inter-zonal road traffic and potential accessibility (using $d_{ij}^{-1.0}$)*

Independent variable	a	b	R^2	F
Traffic generated (O_i) $\log_{10}$ of population-miles	286.4155	-82.3430	0.40	51.26
Traffic attracted (D_j) $\log_{10}$ of population-miles	282.4802	-81.0602	0.40	50.94

In any analysis of the spatial variation in mean haul, the question must arise whether the commodity structure of traffic has an important effect upon the zonal estimates of mean haul. As table 6.11 shows, there are substantial differences in the mean haul for the various commodities and any significant variations in the composition of traffic can therefore be expected to affect the mean haul for the aggregate of all freight. On the other hand, the evidence in chapter 5 indicates that there is no systematic relationship between commodity structure and location with respect to the national market. Therefore, our initial expectation is that the effects of freight composition are spatially random, at least in the context of central versus peripheral locations.

Table 6.11. *Road freight: estimated national mean haul, tonnage and ton-mileage for commodity groups*

Commodity group	Tonnage, million	Ton-mileage, million	Mean haul
Coal and coke	142.1	3,095.3	21.78
Scrap	11.2	318.3	28.51
Steel	45.5	2,393.1	52.58
Oil	57.8	1,845.2	31.92
Transport goods, etc.	8.0	418.5	52.38
Food	306.8	10,754.2	35.05
Chemicals	34.9	1,717.9	49.23
Building materials	444.4	12,026.8	27.06
Other manufactures	150.3	7,394.1	49.20
Other crude	69.4	2,626.3	37.85
Miscellaneous	113.2	3,327.2	29.39
Total	1,383.6	45,916.9	33.17

Note: The total ton-mileage differs slightly from that shown in the Appendix, owing to rounding errors. Mean haul taken from Department of the Environment, 1971, p. 44.

Freight flows and the British economy

Given that a valid estimate of mean haul for each commodity for each of the seventy-eight zones cannot be obtained, owing to the large number of cells in the matrices that are vacant, the following procedure has been used. For each zone, the total tonnage O_i and D_j is known for each commodity. The national mean haul for the respective commodity groups (table 6.11) can be applied to these tonnages to obtain an estimate of ton-miles. In this way a tonnage-weighted estimate of mean haul is derived, which shows what the aggregate mean would be given the actual composition of traffic and assuming the national commodity mean hauls are applied. The difference between this estimated mean haul and the 'actual' mean, therefore, represents the operation of factors other than the commodity mix.

For this exercise, originating road traffic has been used. All flows, both inter-zonal and intra-zonal, have been included. Consideration was given to examining inter-zonal flows only, given that in this case some forty per cent of the variance in mean haul can be explained by the logarithm of potential accessibility, whereas only one quarter is so explained when intra-zonal flows are included. However, there is a serious objection to using only inter-zonal flows for an exercise of this kind. A national figure for the inter-zonal mean haul can readily be calculated for each commodity. But the proportion of all traffic that is inter-zonal varies very widely from zone to zone: for those zones that are geographically very large, most freight is intra-zonal and in these cases the tonnages of inter-zonal traffic for individual commodities are very small and subject to sampling error. Therefore, it seemed wiser to work with the aggregate tonnage of all freight.

The zonal estimates of mean haul assuming the national commodity means applied to the tonnages O_i are shown in the Appendix. The first and very striking fact to note is that whereas the actual mean haul has a coefficient of variation of twenty-seven per cent, the standardised haul has an equivalent variation of only six per cent. In other words, the effect of variations in commodity mix is very small indeed. Thus, the main source of spatial variation in mean haul must lie with factors other than the structure of freight traffic.

The next step was to take the differences between the actual mean haul and the standardised haul and to correlate these differences with the logarithm of potential accessibility ($d_{ij}^{-1.0}$). Differences have been obtained by subtracting the standardised mean from the actual mean. Thus, a positive difference indicates that the actual mean haul is greater than expected, and vice versa. The regression obtained is

$$Y = 68.5548 - 25.0459 \log X \quad R^2 = 0.35$$

clearly indicating that the mean haul is greater than 'expected' where the potential accessibility value is low; the converse holds where potential accessibility is high. If this result is compared with the equivalent R^2 value of 0.26 in table 6.9, it will be seen that elimination of the commodity effects does make for a better explanation of the spatial pattern of mean haul, though we have still accounted for only one-third of the variance of the differences on the basis of potential accessibility.

If it were possible to apply this technique to the inter-zonal traffic alone,

98

it seems likely that a modest improvement could be made in the estimation of zonal mean haul. However, it seems unlikely that much more than half of the variance could be explained by commodity composition and potential accessibility. So we may conclude that, while location is an important variable affecting mean haul and may be the single most important one, it is not an adequate basis on which to forecast zonal means.

The symmetry of O_i and D_j road traffic mean haul

As noted on p. 63 in chapter 5, the fact that at the national level there is a necessary identity of originating and terminating traffic does not mean that a similar equality necessarily holds for each zone. Comparing tonnages O_i and D_j in chapter 5, the intra-zonal element was excluded. However, in comparing the O_i and D_j mean hauls, it is not immediately clear that the intra-zonal element ought to be excluded. Indeed, if we correlate the mean haul for aggregate freight O_i and D_j, the value of R^2 is 0.87 when intra-zonal traffic is included and 0.94 when it is excluded. One might expect the level of correlation to be higher when the intra-zonal element is included, because this provides a component for both the O_i and D_j traffic and has the same mean haul in both cases.

The apparent paradox can be explained quite simply. In all cases, the intra-zonal mean haul is less than the inter-zonal haul. The volume of inter-zonal traffic O_i is not identical to the volume D_j, though as already shown there is a close correspondence (p. 63). Thus, in obtaining the over-all mean haul for O_i and D_j traffic for each zone, the intra-zonal traffic is assigned a somewhat different weight in each case. Consequently, even if the mean haul on inter-zonal traffic O_i were identical to that for the inter-zonal D_j freight, the over-all means would differ. Where, as is usually the case, the inter-zonal hauls do differ, there is no *a priori* basis on which to predict whether the differences will be increased or reduced by including the intra-zonal movements.

Clearly there is a very high level of association between the length of haul outwards from and inwards into the seventy-eight zones. The form of this relationship is interesting. Where Y equals the mean haul on O_i traffic and X the haul on D_j freight, both in miles for inter-zonal traffic.

$$Y = 1.341 + 0.984 X \quad R^2 = 0.94$$

This indicates a slight tendency for originating traffic to travel further than terminating traffic where hauls are short, and vice versa. Since length of haul is partly related to location within the space-economy, we may note a slight tendency for accessible zones to send their products further than the distance from which they draw supplies. By contrast, the remoter regions are apt to send their products to places nearer than those from which they must fetch supplies. Though the tendency is not strongly marked, it is at least consistent with what we would expect. Furthermore, it is consistent with the tendency noted on p. 63 for the major urban areas to consume a larger volume of goods than they produce.

Conclusion

The net conclusion of this chapter is a good deal more optimistic than that of chapter 5. If it is difficult to make useful forecasts of the volume of freight that will be generated by and attracted to zones, it is possible to make considerably greater progress in devising spatial allocation models to assign flows. On the whole, the gravity model performs less well than does linear programming and it looks as though the latter technique may be the more fruitful for further exploration. In general, the more homogeneous the commodity group, the better does the L.P. model perform; conversely, the gravity model works best where the commodity is heterogeneous.

Underlying the work reported in this chapter and the conclusions based on it, however, is the abiding problem of data. It is likely that if the commodity classification were finer, the L.P. formulation would perform better than we have been able to make it work. The same may be true if the geographical zoning were finer. However, perhaps the essential point about the degree of aggregation is indicated by the performance of the L.P. model using the twenty-four main cities. If a substantially finer breakdown of both commodities and geographical areas were available, it would be possible to experiment with the level of aggregation to establish the trade-off between veracity and volume of work involved. At the present time, this is not feasible, even if the resources were available to handle the data and computations, because the data only exist at a very coarse level of aggregation.

Finally, we have shown that there are spatial variations in comparative advantage, in terms of transport, that conform to the postulate in chapter 2 that peripheral areas are at a disadvantage. In the gravity model formulation, peripheral zones do try to avoid some of the penalties of their location as exhibited by high distance friction exponents. The adjustment is incomplete and these regions do also experience a relatively long haul on their freight traffic. Dual values from the L.P. solutions show patterns of shadow prices that can reasonably be interpreted in terms of access from points of origin and which again display elements of the centre/periphery dichotomy.

Chapter 7

MODAL SPLIT

Although the question of modal split is not central to the present study, there
are nevertheless two reasons why it is important to examine variations in the
relative importance of road and rail traffic. The first reason is that in some of
the exercises on which we have reported, attention has been devoted exclusively
to road traffic. If there are systematic variations in the significance of road
traffic relative to rail, then it may be that important errors have been introduced which ought to be corrected. Secondly, in the forecasting of freight
flows it may be necessary or convenient to start by forecasting the aggregate
movement of goods. In this case, the second step is to disaggregate the total
by mode so that estimates may be made of the traffic demands on the road
and railway infrastructure systems. In both contexts, therefore, it is important
to look at the question of modal split to see how far it is possible to go with
the data available to us. However, before doing so, it will be helpful briefly to
review previous work on this general problem.

Previous work

Competition between the railways and public transport facilities on the one
hand, and road use by cars and lorries on the other, is a major debating point
in urban and national transport affairs and consequently some research has been
done to develop models to explain modal choice. Most of this work has been
done on passenger travel to explain choice in terms of travel time, cost and
convenience and the nature of the traveller and his purpose, so that the results
of changes in the transport system can be predicted. Some of this work can and
has been translated into freight terms, considering the nature of the commodity
as well as the transport mode. Warner (1962) used regression and discriminant
analysis to estimate the probability of choosing one mode over another as a
function of travel time, cost and the characteristics of the journey and user. In
Great Britain this work was followed up by Quarmby (1967), who expressed
the problem in terms of the disutility of alternative means of travel. Translating
into freight movement terms, his model was based on the assumption that the
shipper can be expected to choose the mode which minimises disutility. The
relative disutility of one mode compared with another can be expressed as:

$$Z = \alpha_1 x_1 + \alpha_2 x_2 + \ldots + \alpha_n x_n,$$

where $x_1, x_2, \ldots, x_n$ are the characteristics of the transport media and
commodities which give rise to disutility: time taken, cost, regularity, reliability,

101

bulk/value ratio, perishability, length of haul, etc. Discriminant analysis can be used to find the best values of the parameters $(\alpha_1, \alpha_2, \ldots, \alpha_n)$ for explaining the observed choices of a sample of shipments.

Bayliss & Edwards (1970) followed a similar line with regression analyses based on freight consignment survey data, using a binary dependent variable to express the use of road or rail for the journeys. Quandt & Baumol (1966) dealt with both distribution and modal split forecasting in their 'abstract mode' transport demand model (the application of this model to freight traffic was examined in chapter 2). Modes are characterised in terms of cost, speed and frequency, with no reference to the equipment or institutional structure of the service. A shipper's decision to ship and his choice of mode are considered to depend on the absolute performance level of the 'best' mode on each criterion, and also the performance level of each mode on each criterion relative to that of the best mode. Since the model is couched in terms of 'abstract' criteria, in order to forecast the use of new or improved modes, the performance characteristics of the mode are merely inserted into the estimating formula, as was detailed in chapter 2.

These approaches require considerable quantities of data on the cost and quality characteristics of the transport modes and are designed to deal with individual consignments. They have not been developed in the context of the much cruder data available to us. On the freight side in particular, we have no information on the size of individual consignments and the regularity of dispatch. The measurement of distance is very crude – the distance between centroids – and of course with only 13 commodity classes we are dealing with heterogeneous goods. The only study of which we are aware that attempts an analysis of modal split with the kind of data we have is that included in the Ministry of Transport's Transport Costs Model. For this exercise, as we understand, the proportion of road freight to rail freight was established for each commodity by fairly broad distance zones. In the T.C.M. it was assumed that these ratios would hold in the future. Therefore, once the total freight flow between a pair of zones had been estimated, the allocation to road and rail was based on the proportion appropriate to the inter-zonal (or intra-zonal) distance class into which the particular flow fell.

Possibly, there is not much more one can do with such data. It would not be legitimate, in any case, to transfer models developed for individual decision makers to examine or predict zonal aggregate behaviour, especially when the zones are so varied in size and character. Furthermore, it would be necessary to have detailed information on the cost of transport, by mode and commodity, as well as data on the journey time and other characteristics of the transport systems. Though we have some information on road transport costs (chapter 9), they are not adequate for an analysis of modal split and are not complemented by comparable data on rail costs.

Overall, therefore, it is apparent that previous work in this field is not very relevant to the problem in hand. It also becomes apparent that only quite elementary analysis is feasible. Two propositions may be made, of which the second is particularly important. First, the bulkier a commodity is relative to its value, the higher the proportion that is likely to move by rail. Secondly, the cost structures of road and rail haulage give the former a competitive edge for short hauls and the latter for long hauls. Thus, it is to be expected that on the longer hauls rail freight will be relatively more important, and vice versa. Both propositions are examined in the following pages.

Evidence on modal split

Table 7.1 shows that rail traffic is only significant for a limited number of commodity classes. Apart from the fact that limestone and iron ore move exclusively by rail, the railways are significant for only coal and coke, scrap and steel. In all other cases, rail traffic accounts for less than one-fifth of the

Table 7.1. *Distribution of freight between road and rail, individual commodities, million tons*

Commodity	Road	Rail	Total
Coal and coke	142.1	152.8	294.9
Iron ore	–	19.8	19.8
Limestone	–	3.3	3.3
Scrap	11.2	5.9	17.1
Steel	45.5	17.3	62.8
Oil	57.8	8.0	65.8
Transport goods, etc.	8.0	0.8	8.8
Food	306.8	6.2	313.0
Chemicals	34.9	4.6	39.5
Building materials	444.4	17.1	461.5
Other manufactures	150.3	1.8	152.1
Other crude	69.4	7.3	76.7
Miscellaneous	113.2	0.9	114.1
Total	1,383.6	245.8	1,629.4

tonnage. Moreover, it can be seen that in general the railways are more important for the commodities which are bulky in relation to their value, and conversely. Thus, one of our initial expectations is confirmed.

The next consideration harks back to chapter 6, in which it was shown that both the gravity model β values and zonal mean haul vary spatially. In that chapter, it was also shown that these variations do manifest some relationship to location within the country, the zones with long mean hauls tending to be peripheral. If rail traffic has a competitive advantage over road traffic for long hauls, then we would expect rail traffic to be relatively more important in the movement of goods for peripheral zones than for central ones. To test this expectation, road freight tonnage was expressed as a percentage of the total for each zone, originating and terminating traffic being treated separately. These percentages were taken as the dependent variable and were regressed first on population-miles and then on the mean haul for road traffic alone. It seemed wise to conduct two experiments, since although there is some relationship

103

between mean haul and population-miles, it is not very great. The linear regressions were of the form:

$$Y = a + b \log X_1,$$

and

$$Y = a + b X_2,$$

where Y = the percentage of all freight moving by road
X_1 = population-miles ($d_{ij}^{-1.0}$),
X_2 = mean haul for road traffic, intra-zonal freight being included.

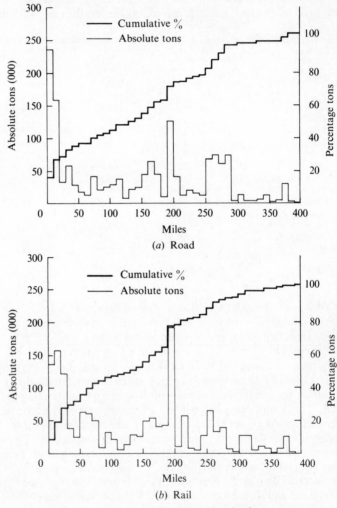

(a) Road

(b) Rail

Fig. 7.1. Modal split: coal and coke

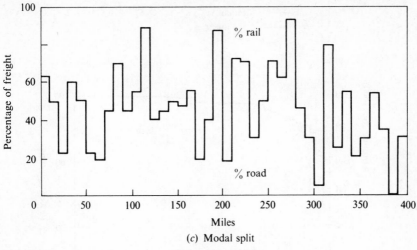

(*c*) Modal split

Fig. 7.1. (cont.)

In no case did the value of R^2 rise above 0.06, indicating very clearly that no significant relationship exists between the variables.

Given the non-significance of the equations, it is clear that there is no tendency for peripheral zones, or zones with long hauls on road freight, to use rail services more than the centrally located zones. Nor is there any evidence pointing to the opposite conclusion. A clear implication of this finding is that the analysis in chapter 6 of the spatial variation in the characteristics of road traffic is in no way suspect on account of the omission of rail freight.

The absence of any spatial regularity in the relative importance of road and rail traffic may seem surprising. However, this finding is consistent with the evidence in chapter 5. For estimating the generation and attraction of freight, it was found that a very low level of performance could be achieved for rail freight compared with road traffic. Perhaps the main reason for this is the nature of rail freight, consisting of a limited number of commodities with very specific sources of supply and destinations. Thus, we may infer that the proportion of traffic carried by the railways, both O_i and D_j, varies spatially according to the location of quite specific activities, and not in response to the general location of a zone within the country.

To pursue this matter further, the trip length and modal structure of four commodities will be examined in detail. Similar analyses were in fact performed for the other commodity groups and the selected four – coal and coke, steel, transport goods and food – have been chosen as representative of the various categories of freight. The absolute and cumulative percentage trip-length frequency distributions by 10-mile distance intervals for each commodity by road and rail separately are shown in figures 7.1 to 7.4. Also shown are graphs of the percentage split between road and rail against distance, again by 10-mile intervals.

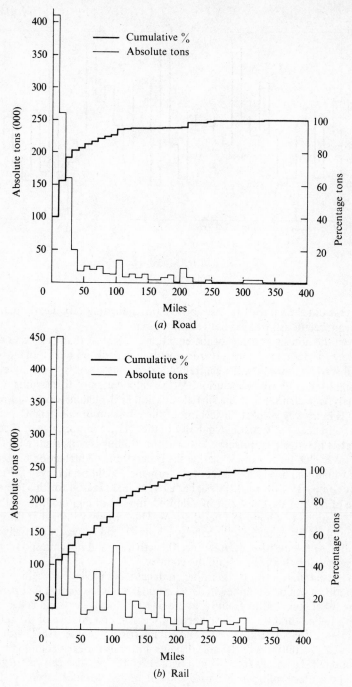

Fig. 7.2. Modal split: steel

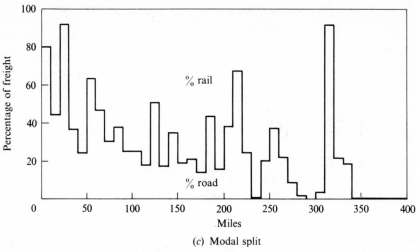

(*c*) Modal split

Fig. 7.2. (cont.)

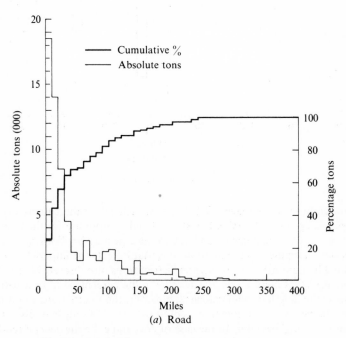

(*a*) Road

Fig. 7.3. Modal split: transport vehicles and equipment

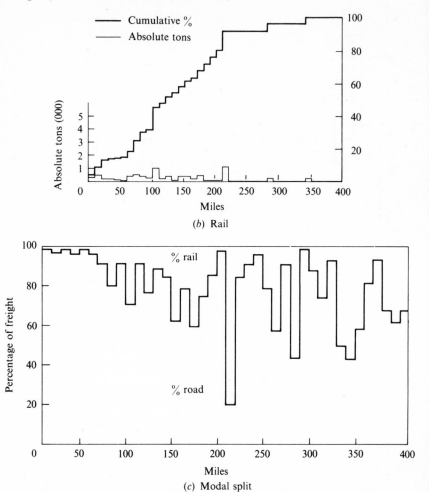

(*b*) Rail

(*c*) Modal split

Fig. 7.3. (cont.)

To obtain these frequency distributions, each cell of the commodity matrix has been assigned to the 10-mile distance band appropriate for the inter-centroid distance, or the intra-zonal mean haul, as the case may be. Consequently, there is a certain lumpiness in the distributions that reflects the imperfect data obtained by compressing continuous distributions into discrete blocks. Nevertheless, note that there is a general increase in the percentage of freight travelling by rail as the length of haul increases; however, the fluctuations about this general tendency are so violent, except in the case of food, as to discourage any curve-fitting exercises. In the case of coal and coke, the general tendency is very indistinct, partly because of the unusual distributions of both road and rail

108

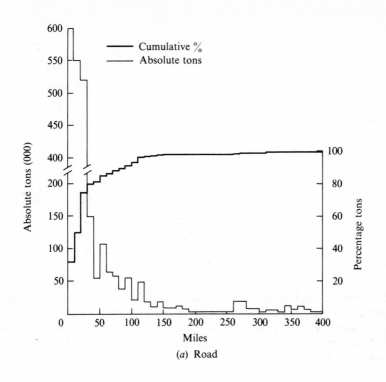

(a) Road

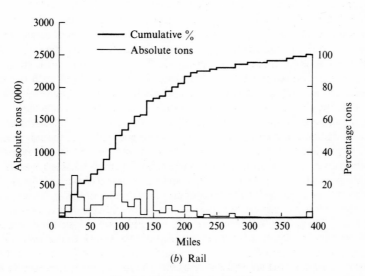

(b) Rail

Fig. 7.4. Modal split: foodstuffs

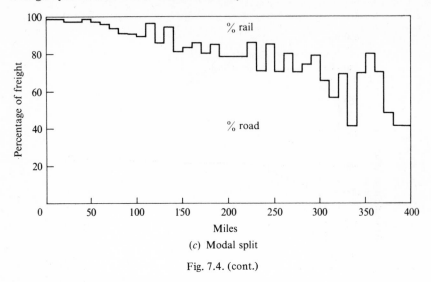

(c) Modal split

Fig. 7.4. (cont.)

tonnage with length of haul. Neither has the usual negative Pareto form; instead there are strong peaks at 10–20 miles and approximately 200–300 miles. The nearly diagonal form of the cumulative percentage curves, with strong kinks at the peaks, bears further witness to this. For steel traffic, the railways clearly increase their share of the traffic as the length of haul becomes greater, while road movement dominates rail for the very short movements. Road traffic totally dominates in the movement of transport goods, rail movement being scattered along the whole spectrum of length of haul up to 400 miles. Apart from some small very long-distance movements, road movement clearly predominates in the transport of food. This domination does decrease fairly regularly with distance, as we would expect. The fluctuations in the percentage split curve are less for this commodity than for the other three, with the large fluctuations occurring at greater distances, where the small volume of deliveries can cause big changes in modal split between 10-mile distance intervals.

From these examples it is abundantly clear that no simple relationship between distance and mode of transport can be postulated. Although there is some suggestion that the proportion of goods moving by rail increases with greater distance, the pattern is far too irregular to permit reliable generalisation by fitting curves. In this context, the commodity group coal and coke is particularly striking, as for this commodity class the relationship between modal split and distance is especially fuzzy, when one would expect it to be particularly clear. Altogether, therefore, we may conclude two things. First, because there appears to be only a weak relationship of modal split to distance, no serious bias is introduced into the other parts of our study. In the second place, it does not appear that anything very useful on modal split can in fact be done with the data in the form we have.

Chapter 8

THE TRANSPORT COSTS OF
MANUFACTURING INDUSTRY, 1963

Previous chapters have shown that there are detectable spatial variations in transport inputs both at the aggregate level and also for commodity classes. It is, therefore, worth asking whether these variations are systematic enough, and of sufficient magnitude, to be evident in the cost-structure of enterprises. For this purpose we turn to the 1963 census of manufacturing industries, the latest full census available when this study was being made. It must be kept in mind that manufacturing industry accounts for only part of all freight movements and also that the standard industrial classification cannot be matched with the commodity classes available for freight movements and used in previous chapters.

The 1963 census was the first to include questions designed to elicit information on total expenditure by firms on transport of all kinds. Previous censuses confined attention to the amount firms spent in purchasing transport services from other firms – the railways, road haulage contractors, etc. Unfortunately for our purposes, full details on all transport outlays, including transport on own account (mainly C-licensed road vehicles), were obtained only for the larger firms, i.e. those employing twenty-five workers or more. For the smaller firms, information was sought on the outlays to purchase transport services, own-account transport being ignored. Furthermore, in the regional analysis of the census data, for both larger and smaller firms, the only transport costs identified are for purchased services.

In aggregate, the larger firms spent £568.6 million on transport of all kinds, of which £341.0 million were payments to other organisations. If the same ratio applies to the smaller firms, it means that the regional tables of the census exclude forty per cent of expenditure on transport. This is a serious deficiency in the data and the first consideration was whether a valid estimating procedure could be devised to provide the full picture. At the national level, Edwards (1970a) used the following procedure. For each industry, three items of information are available:

Larger firms:	purchased transport
	own-account transport
Smaller firms:	purchased transport

Taking the ratio of purchased to own-account transport for the larger firms and applying this to the purchased transport of the smaller firms allows an estimate to be made of the latter's expenditure on own-account services. This estimating procedure is not available at the regional level, since the only data are for

111

purchased services by all firms. Therefore the only way to obtain regional estimates of the total outlay would be to apply to each industry its national ratio of purchased to own-account outlays. This would effectively mean that forty per cent of the total expenditure was assumed to vary in direct proportion to the other sixty per cent. Since we are interested in spatial variations in the proportionate incidence of transport costs, this would not be very helpful.

Therefore, examination of the spatial variation in transport costs must be relatively crude, depending on data for purchased transport services only. Of the various available measures of outputs and other inputs, net output is the most appropriate to use. Net output represents the value added to materials in the course of manufacture and is obtained by deducting 'from the gross output the costs of purchases adjusted for stock changes, payments for work given out to other firms, and payments to other organisations for transport'. A convenient measure of the relative importance of transport is, therefore, given by the value of purchased transport services as a percentage of net output. This percentage can be obtained for each Main Order manufacturing industry (Orders III to XVI) for each region in Great Britain (Board of Trade, 1970, vol. 133, summary tables 21–48A).

Two industries have been omitted from the analysis, namely Orders VII and XI. The former is shipbuilding and marine engineering and is omitted for a reason that highlights one of the weaknesses of the census data for our purposes. The cost of transport recorded for census purposes is outward from the firm, representing transport costs on the delivery of completed or part-completed manufactures. Costs of inward delivery are not taken into account, except to the extent that they are included in the cost of own-account transport. As the end-product of shipyards sail, overland transport costs on delivery are negligible for this industry. Order XI is leather, leather goods and fur manufacture. With a total net output for the United Kingdom of £59.3 million in 1963, this is by far the smallest of the Main Order manufacturing industries. It is also one that displays marked peculiarities in the demand for transport (Chisholm, 1971a), having an astonishingly short haul on its goods. For these two reasons it seems appropriate to eliminate this industry from the analysis.

Table 8.1 shows the ten standard regions of Great Britain and the incidence of transport costs in relation to net output. The first column shows that there is virtually a twofold range, from 2.61 per cent (West Midland) to 5.01 per cent (North). However, this variation may arise from four sources:

(1) differences in the industrial mix of the regions, at the Main Order level,
(2) variable regional incidence in transport costs for each Main Order industry,
(3) variable proportion of total transport cost represented by payments to other organisations,
(4) heterogeneity of the Main Order classes.

There are no adequate means for testing the last possibility, given that for many of the Minimum List industries regional data are incomplete. However, it is possible to examine the first two sources of variation and to throw some light on the third.

112

Table 8.1. *Great Britain, 1963: cost of purchased transport in relation to net output, manufacturing industries (Orders III–XVI but excluding VII and XI)*

Region	1 Weighted mean[1]	2 Unweighted mean[2]	3 Imputed at national level for each industry[3]	4 Net output, £m
	Transport cost as % of net output			
North	5.01	4.96	3.74	522.6
Yorkshire and Humberside	3.95	4.12	3.41	1,010.8
East Midland	3.05	3.90	2.97	675.6
East Anglia	5.12	5.89	3.50	203.1
South East	2.84	3.48	2.96	3,245.3
South West	3.20	3.92	2.99	458.9
West Midland	2.61	2.98	3.07	1,446.9
North West	3.62	3.92	3.30	1,616.1
Scotland	4.15	4.69	3.45	804.3
Wales[4]	4.67	4.56	4.08	432.7
Great Britain	3.20	4.24	3.20	10,416.3

Source: Board of Trade, 1970, vols 131 and 133.
[1] As recorded in the census.
[2] Mean of Main Order classes unweighted.
[3] National % for each Main Order industry applied to net output in the region.
[4] At the Main Order level, no data are published for industries VII and VIII. Therefore column 1 includes VII, whereas in columns 2 and 3 VIII is excluded and column 4 includes both VII and VIII.

Column 2 in table 8.1 shows the unweighted mean cost of transport in the regions. The figures have been derived as the arithmetic mean of the percentage cost of transport for each of the Main Order industries; they therefore take no account of the relative importance of each industry in the respective regions. The incidence of transport costs is raised by about one-third, from 3.20 to 4.24 per cent compared with column 1. In eight out of the ten regions, the unweighted mean is higher than the weighted mean, usually by a substantial amount. The two exceptions are the Northern Region and Wales, but in both cases the difference is relatively small. The implication is that in most regions there is a tendency for those industries with a low incidence of transport costs to be relatively more important than those with high transport costs; as this reflects the national industrial structure, this finding is not especially surprising. Much more interesting is the fact that the variation from region to region shown in columns 1 and 2 hardly differs. The coefficients of variation are respectively twenty-four and nineteen per cent, and a rank order correlation yields an r_s of 0.985, which is significant at the ninety-nine per cent level. Thus, the industries that are located in, for example, East Anglia all tend to have higher transport costs than the same Main Order industries located in any other part of the country. Transport costs, as shown by column 2 in the table, are lower in the West Midlands and the

113

South East than anywhere else, followed by the North West and South West. The Northern region, Scotland and Wales all suffer from relatively high transport costs for all the Main Order industries.

While there is some suggestion that the central regions enjoy economies in transport compared with the peripheral regions, there is also the absolute size of the manufacturing sector in each region to take into account. If the regions are ranked according to the value of net output in Orders III to XVI (but excluding VII and XI), there is an inverse correlation with the arithmetic mean percentage of transport cost (column 2 in table 8.1) which yields a value of r_s of -0.68; this is acceptable at the ninety-five per cent level, but not ninety-nine per cent. This suggests very strongly that it is the magnitude of the industrial economy of a region that affects the level of transport costs at least as much as the location. Plotting the data graphically, there is a strong hint that the relationship is non-linear (figure 8.1). Once a region has achieved a net output in manufacturing of the order of £1,000 million, most of the possible savings in transport appear to have been achieved. The two regions that have a substantially lower manufacturing net output, but also display a low incidence of transport costs, are the East Midlands and the South West. Most of the latter's manufacturing is in the east of the region; therefore, both regions may be classed as reasonably 'central' in the space-economy of the country. Thus, it is the Northern Region, Scotland and Wales that appear to be at a disadvantage on account of the combination of a modest manufacturing base and peripheral location, while East Anglia is exceptional owing to its unusually small industrial base.

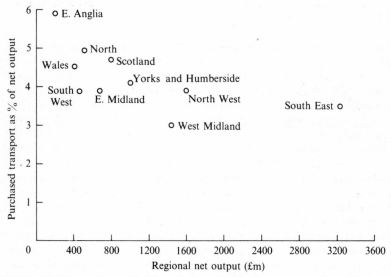

Fig. 8.1. Manufacturing industries, 1963. Unweighted regional mean incidence of purchased transport costs related to the regional value of net output. Orders III to XVI but excluding VII and XI.

The transport costs of manufacturing industry, 1963

A further way in which the spatial incidence of transport costs may be examined with the aid of the census data is to ask the question: if for each industry the incidence of transport costs in each region were the same as the national average for that industry, what impact would this have on the spatial pattern of transport costs? Column 3, table 8.1, gives the answer, derived by applying to the net output of each Main Order industry in each region the national (United Kingdom) proportionate incidence of transport costs. The effect of this operation is to keep the total outlay on transport the same as that recorded in the census, but to redistribute it among the regions. The rank order of the ·egions is very little affected; for the pairwise correlations of columns 1/3 and 2/3 the respective values of r_s are 0.88 and 0.82, both significant at the ninety-nine per cent level. On the other hand, regional disparities in the impact of transport costs are substantially reduced, the coefficient of variation in column 3 being only eleven per cent.

The fact that when the percentage expenditure on transport for each industry is standardised at the national level the regional coefficient of variation falls to eleven per cent is highly suggestive. It will be recalled that the regional coefficient of variation for the weighted data is twenty-four per cent and for the unweighted data is nineteen per cent. This implies that about half the regional variation in the overall cost of transport arises from differences in the industry mix and that the other half is due to the variable incidence of transport costs on each industry class.

It will be recalled that payments to other firms for transport services amount to sixty per cent of the total outlay of larger firms. In the absence of adequate regional data, a direct check on the magnitude of potential error introduced by using data for purchased transport only is not feasible. However, we may suppose that the error is likely to be greater for industries in which the purchased transport accounts for a lesser proportion of the total, and vice versa. Furthermore, we might expect the regional variation in (purchased) transport as a percentage of net output to be the greatest for industries where the purchased transport is least important. The basis for this last proposition is that purchased and non-purchased transport are substitutable. For an industry which in national terms uses comparatively little purchased transport the possibilities of substitution imply that considerable variations may occur regionally in the proportion of purchased transport to net output.

To test this possibility, the incidence of purchased transport relative to net output was calculated for each Main Order industry in each region. Hence, the unweighted coefficient of variation was derived for each industry. This measure of the spatial variability of transport costs for an industry can be compared with the proportion that purchased transport is to all transport outlays at the national level. It is to be expected that the coefficient of variation will be greater the lower is the overall importance of purchased transport, on account of substitution between purchased and own-account transport. In fact, as figure 8.2 shows, the relationship is exactly opposite to that which is expected. The industries clearly fall into two groups and yet there are no very obvious elements so linking them. Thus, for those industries appearing in the top right of the diagram it is unlikely that own-account transport can materially affect the spatial variation in the incidence of transport costs. On the other hand, industries towards the lower left include Orders III and XIV, which are characterised by comparatively short hauls on local delivery (Chisholm, 1971a). These users of transport, because they serve mainly local markets, are unlikely to have a highly variable proportion of transport expenditure on own account.

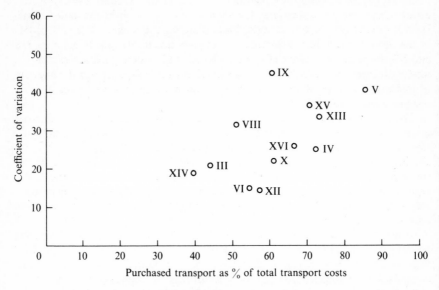

Fig. 8.2. Manufacturing industries, 1963. Unweighted industry coefficients of regional variation in the proportion of purchased transport costs to net output related to the national proportion of total transport costs represented by purchased transport services. Orders III to XVI but excluding VII and XI.

On the face of it, therefore, the omission of own-account transport from the regional analysis is not likely to be a major defect, though clearly the results must be treated with due caution.

Overall, the evidence from the 1963 census indicates that regional variations in transport costs are not very great. Though there is some suggestion that central regions are advantageously placed relative to the peripheral ones, it is equally plausible to suggest that it is the size of the industrial base that is the most important factor. As the regional data on transport costs are unsatisfactory in several ways, this conclusion must be treated with caution. However, it is consistent with the evidence that the spatial variation in the volume of transport, especially the affect on mean haul, is only loosely related to the question of location with respect to main centres of population: most of the standard regions contain at least one major industrial complex. Furthermore, the absence of evidence on spatial variations in transport costs is also consistent with evidence on the structure of freight charges, examined in chapter 9.

116

DISTANCE AND THE COST OF TRANSPORT*

Thus far, our analysis has been based on the assumption that space is adequately represented by distance in miles. In most of the exercises distance was untransformed, except in the experiments with the gravity model, where various distance exponents were used. Clearly it is desirable to examine this assumption that transport cost is a monotonic function of distance. Such an examination may be undertaken for two reasons. The first is that, if realistic estimates of transport costs can be obtained for consignments sent over various distances, it may be desirable to re-formulate both gravity and linear programming models to express distance in monetary cost terms. For this to be warranted, detailed and reliable transport cost estimates would be necessary. An alternative reason for examining transport costs is to examine the form of the relationship between distance and transport costs as an aid in the interpretation of results obtained using the very general assumption that transport costs are a monotonic function of distance. It is in the spirit of the second, not the first, objective that the following discussion is couched.

The greater the distance a consignment is sent, the greater is the cost of shipment, if all other things are equal. At the simplest, one may assume with Weber that costs are directly proportional to distance, or that some degree of tapering in freight rates occurs. But our main interest attaches to the relationship between terminal costs and movement costs, since the larger the former are relative to the latter, the less is the significance of increments in distance. Our problem is to measure the relationship between terminal charges and movement charges. Most of the data in this field are based on a rather different concept, namely the distinction between vehicle standing costs and running costs; this is clearly important from the operators' point of view and is valid in accountancy terms, but is not very closely related to the distinction between terminal costs and movement costs. For our purpose the requirement is to obtain data on actual freight charges in relation to distance.

1962 saw the passage of a Transport Act that freed the railways from many of their common carrier obligations, enabling them to cease publishing freight rates and to enter into special contracts with individual consignors. Consequently, data on rail charges in the mid-1960s are not readily available and the rates for earlier years are likely to give a misleading impression. On the other hand, a

---

* The greater part of this chapter was first published in M. Chisholm and G. Manners (eds.), *Spatial Policy Problems of the British Economy*, C.U.P., 1971. The material is reproduced with the permission of the Syndics of Cambridge University Press.

117

certain amount of information has been collected for road transport and it is
on this mode that the discussion will mainly focus. First, however, we will
review some recent findings that are relevant to the question in hand.

Bayliss & Edwards (1970) obtained a national sample of freight consignments
made by industries and their data include information on the charges for consignments sent by rail and professional road haulier. They were able to 'explain'
a very high proportion of the variation in charges by factors such as length of
haul, the weight of consignment and the volume of freight annually shipped
from the establishment by the mode (road or rail) in question. Estimating
equations were obtained for the two modes separately and 'in both models
consignment weight was by far the most significant variable, explaining more
than four-fifths of the variation in charges' (Bayliss & Edwards, 1970, p. 159).
Length of haul proved to be comparatively unimportant as an independent
factor. Data from the West Cumberland survey provide dramatic support for the
proposition that consignment size is an important factor affecting unit charges,
as is shown in table 9.1.

Table 9.1. *West Cumberland Survey: average cost of consignments by public transport*

Weight category of consignment	Cost per ton			Cost per ton-mile
	£.	s.	d.	d.
up to 22 lbs	10	4	11	217
22–56 lbs	3	18	7	69
56–112 lbs	2	18	8	48
112–560 lbs	1	10	1	36
560–2240 lbs		13	7	12
1–2 tons		15	0	11
2–3 tons		10	5	8
3–4 tons		7	8	7
4–5 tons		5	0	5
5–7 tons		5	0	5
7–10 tons		3	7	5
10–15 tons		3	1	4
15–20 tons		4	4	4
20 tons and over		4	3	4

Source: Edwards, 1967, *Statistical Appendix*, p. 26.

This interpretation cannot be taken it face value, since the size of consignment is correlated with length of haul and also with the volume of traffic. This
is clearly demonstrated by evidence from the Severnside study (Edwards, 1970b).
Using Spearman's rank correlation for consignments of textiles, fabrics and
yarns, clothes and hoisery, an r_s value of 0.83, significant at the ninety-nine per
cent level, was obtained between volume of flow and consignment size. Consequently the Bayliss and Edwards results are not very helpful in the present
context, though they do suggest that it is the size of consignment, its nature
and the volume of traffic that are the main determinants of cost, rather than

distance as such. However, if there is a high correlation between these factors and distance, then the distance variable is a useful one to use.

Deakin & Seward (1969) also carried out a national sample survey, this time of hauliers. Road haulage firms were asked the average charge they made and the mean distance of haul for each of thirty-four commodity groups. The unpublished summary data have kindly been made available to us. For some commodities there were fewer than fifteen usable observations and in a few cases where data were more numerous no meaningful linear expressions could be computed. Nevertheless, for twenty-three commodities regressions of the form

$$Y = a + bX,$$

where Y equals the transport charge in shillings per ton and X equals length of haul in miles, yielded results acceptable at the ninety-nine per cent level. The results are set out in table 9.2.

Table 9.2. *Great Britain, 1966: road haulage charges in shillings per ton related to length of haul*

Commodity	Unused observa-tions	N	a	b	$a \div b$	R^2
Electrical and non-electrical machinery and transport equipment	0	29	27.736	0.161	172	0.45
'Other' chemicals and plastics	2	32	21.658	0.187	116	0.61
Metal manufactures	5	28	21.186	0.185	115	0.69
Iron and steel, finished and semi-finished	2	48	21.100	0.147	144	0.61
Wood, timber and cork	2	37	19.560	0.155	126	0.63
Mixed loads	3	34	19.499	0.200	97	0.64
'Other' manufactured goods	2	36	19.133	0.209	92	0.55
'Other' foods and tobacco	7	29	17.888	0.216	83	0.64
Petroleum and petroleum products: gas	0	16	17.676	0.201	88	0.67
Flour	0	18	16.474	0.148	111	0.87
Fresh fruit, vegetables, nuts and flowers	7	19	16.259	0.241	67	0.80
Textiles (fibres and waste)	1	15	15.817	0.269	59	0.88
Oilseeds, etc.	1	26	15.354	0.190	81	0.82
Beverages	2	17	14.420	0.247	58	0.91
Fertilisers	0	44	13.968	0.144	97	0.81
Iron ore and scrap iron	3	29	13.479	0.125	108	0.63
Animal feeding stuffs	4	46	13.037	0.219	60	0.66
Cement	0	32	12.543	0.142	88	0.75
Lime	0	20	11.743	0.185	63	0.73
Cereals	4	32	10.434	0.231	45	0.82
Coal and coke	4	40	8.973	0.223	49	0.63
Building materials	10	64	7.086	0.251	28	0.73
Crude minerals other than ore	1	48	2.891	0.259	11	0.77

Source: Deakin & Seward (1970).
All the regressions are acceptable at the 99 per cent level. See text.

Freight flows and the British economy

The Deakin and Seward data were obtained from a sample of firms and there is considerable variation in the volume of traffic handled by them; no weighting has been used in calculating the regressions. Some of the observations appeared to be clearly maverick, being much too high or low a charge for the given distance or being an exceptional length of haul; these maverick data have been excluded from the calculations. Table 9.2 shows the number of eliminated observations and the number (N) used for the regressions. Consequently the coefficients shown in table 9.2 must be treated with care and regarded only as giving orders of magnitude for the listed commodities.

The intercept of the regressions (a in table 9.2) may be interpreted as the terminal charge, which is invariant with distance; the slope, or b coefficient, indicates the increment in cost in shillings per ton for each mile of additional haul. Dividing a by b shows the distance a consignment must be sent to incur a movement cost equal to the terminal cost. The unweighted average coefficients and the median values for the twenty-three commodities may be compared:

	a	b	a ÷ b
Unweighted mean	15.561	0.197	79
Median	16.817	0.200	79

Using the median coefficients as a synthetic estimating equation, Chisholm (1971a) has shown that for 1962 the proportion of road freight cost that is invariant with the length of haul — the terminal element — was between sixty-nine and seventy-five per cent. In that year, according to the Ministry of Transport survey, the mean length of haul on road freight was 26.9 miles. The 1970 *Annual Abstract of Statistics* shows the mean haul to have been 27.9 miles in 1964, whereas our own estimate from the 1964 matrix of road flows is 33.2 (see pp. 35 and 97). At a mean haul of 33.2 miles, 70.4 per cent of road freight charges may be counted as fixed and unrelated to the length of haul.

We have already seen that there are considerable variations in the length of haul in various parts of the country. If it may be assumed that the charge per ton (in shillings), in all parts of the country, varies as $16.817 + 0.200X$, X being the mean haul in miles, it is possible to estimate the proportion of the total freight bill in each zone that is invariable with length of haul. Table 9.3 shows the mean and standard deviation values for the seventy-eight zones, expressing terminal charges as a percentage of total charges at the respective lengths of haul. (Note that in this table the means of 32 and 31.5 miles are the unweighted means derived from the seventy-eight observations, and these differ slightly from the overall estimated mean haul.) It will be seen from table 9.3 that the vast majority of zones lie in the range 62–77 per cent; in almost all cases, therefore, terminal charges represent at least two-thirds of freight charges. The zone with the shortest mean haul is number 16, with Sheffield as its centroid, where the proportion of terminal charges is eighty per cent for the O_i traffic and seventy-nine for the D_j freight. At the other extreme, zone 68 centred on Aberdeen has a mean haul some 3.5 times greater; terminal charges, nevertheless, still amount to fifty-three and fifty-four per cent respectively for the originating and terminating movements.

120

Table 9.3. *Road freight: terminal charges as a proportion of total charges*

	Standard devia- tion about mean haul	Haul in miles	Terminal charges as a percentage of total charges
Originating traffic (O_i)	+2	49.5	61.5
	+1	41	66
	$\bar{x}$	32	71
	−1	23	77
Terminating traffic (D_j)	+2	48	62
	+1	40	67
	$\bar{x}$	31.5	71.5
	−1	23	77

Another way of expressing the effect of the pattern of freight charges is to compare the means, standard deviations and coefficients of variation for mean haul and estimated charge, for the seventy-eight zones. This is done in table 9.4. It will be seen that the coefficient of variation for the unit charge is only about one quarter of the equivalent coefficient for mean haul. On the assumptions made, we can say that some two-thirds of the zones experience a unit charge for their freight that diverges from the mean charge by less than ±8 per cent.

Table 9.4. *Road haulage: zonal variation in mean haul and unit charge*

	Mean	Standard deviation	Coefficient of variation
Mean haul, miles			
O_i	32.1	8.7	27.1
D_j	31.5	8.3	26.3
Unit charge, shillings per ton			
O_i	21.9	1.7	7.8
D_j	21.8	1.6	7.3

It must be accepted that the estimates of transport charges here presented are open to challenge; they must not be treated as firm 'fact', but rather as the best estimate currently available. However, the magnitude of terminal charges relative to movement charges appears to be so great that even if the estimates are seriously in error the following conclusion will remain valid. The structure of freight charges by road is such that, over the relatively short hauls which characterise this country, the spatial variation in transport costs to the user is substantially less than the variation in the volume of transport inputs as measured by length of haul. In this way, therefore, the significance of location is materially reduced. Undoubtedly this is one of the reasons why the 1963 census of manufacturing industries shows so little regional variation in the significance of transport (see chapter 8).

Freight flows and the British economy

The reader will also recognise that the findings reported in this chapter have a bearing on the role that location theorists assign to transport costs. It is a clear implication that, at least for a country the size of Britain, the locational significance of transport costs is much less than formal theory would have one believe.

The final implication to be drawn from this analysis of freight transport charges is that the assumption employed in earlier chapters, that transport costs are a monotonic function of distance, is valid. Furthermore, although terminal charges make up a large proportion of the total, movement charges are a linear function of distance. This is of considerable importance, because it means that in both the gravity model and linear programming approaches it is justified to express distance in terms of miles. Consequently, there is no need to enter any reservations on the previously reported results in view of the evidence adduced in this chapter.

Chapter 10

SUMMARY AND CONCLUSIONS

In this final chapter, we shall deal with four issues. The first is to summarise the main conclusions that have emerged from the empirical work reported in preceding chapters. More important, perhaps, is the second task, to interpret these findings and especially to point out the implications for policy decisions by government. Thirdly, throughout the whole of our work we have been acutely conscious of the problems imposed by the nature of the data available: while it is fatuous to issue a general plea for more and better information, it is, nevertheless, the case that a limited number of improvements might well yield a handsome return in better models and hence a greater forecasting capability. Finally, the question arises as to where future enquiry may be best directed? In this vein, the work reported in the present book should be regarded as explorations that point to the need for further enquiry in some directions at least, rather than providing a definite solution of all problems.

Main findings

An initial question examined in chapter 3 is whether, for the purpose in hand, the British economy can be treated as a closed one. Although in monetary terms it would be patently absurd to assume a closed economy with no external transactions, the same conclusion does not apply to the volume of freight measured in tonnage terms. Oversea trade is equivalent in volume to about twelve per cent of inland freight movements, for Great Britain as a whole. Though not a high proportion, the magnitude is sufficient to raise the possibility that internal movements may be affected, especially for port zones. Further examination of the available data indicates that the impact of foreign trade on the inland freight movements of ports is too small to be detected. As an initial proposition, therefore, it was assumed that the economy is closed and in subsequent work no evidence came to light to suggest that this was a mistake.

Examination of the pattern of freight generation and attraction revealed that resident population and employment are moderately good predictors of zonal freight volumes at the aggregate level, for road alone and for road plus rail traffic. Retail turnover proved to perform less well as a measure of demand. However, the three gross variables proved gravely inadequate for prediction of the volumes of the individual commodity groups. Disaggregating the employment variable into the twenty-four S.I.C. classes, considerable success was achieved for some commodities. On the generation side, ninety per cent or more of the variance

123

is explained in the case of coal and coke, transport equipment and other manufactures; in the case of attraction, only for steel was this degree of success achieved. Thus, with the presently available data, the basis for forecasting the zonal generation and attraction of freight by the use of exogenous variables is not very strong for road and road plus rail; with rail traffic only it is clear that other techniques for estimating freight volumes must be sought. On the other hand, clear evidence emerged that the aggregate volume of road freight is in no way related to the location of a zone with reference to central or peripheral parts of the space-economy. Remote zones do not seek to avoid some of the penalties of their location by engaging in activities that require a small volume of freight per person.

If only moderate success was achieved in finding ways to estimate the volume of freight originating in and destined to zones, attempts at modelling flows given the O_i and D_j volumes proved much more encouraging. The gravity model formulation, applied to aggregate road freight flows, yielded a level of explanation in the order of eighty per cent, which is about the same as the level attained in estimating volumes generated and attracted, using the gross variables of population and employment. Disaggregated to commodities, the gravity model performed less well. Doubt was also cast on the utility of the gravity model by the very large range in distance exponents obtained for the seventy-eight zones, indicating that a single model for the whole country must lead to serious over- and under-estimation for individual zones. On the other hand, linear programming solutions for the commodities provided a better fit with reality, though still not sufficiently good to allow confident predictions to be made. Success came with restricting the analysis to the twenty-four main urban areas: only for one commodity, other manufactures, did the relationship between the L.P. solution and actual flows fall below ninety per cent and in the majority of cases the level of explanation achieved was over ninety-six per cent. Other main conclusions are that the behaviour of freight flows is affected by location within the space-economy. Peripheral zones do have a greater distance-friction, as measured by the gravity model β values, than central zones. However, this adjustment is incomplete and as a consequence the peripheral zones also have a longer mean haul on freight movements than do the more accessible ones. These findings are consistent with the evidence of the dual values (shadow prices) obtained from the L.P. solutions for six commodity classes.

The evidence from the gravity model and L.P. analyses, including the spatial variation in the β values, shadow prices and mean haul on freight, shows that there are spatial variations in the transport inputs required to conduct economic activities, measuring inputs in ton-miles. Attention was therefore turned to the latest available census of manufacturing, to see whether these variations can be detected in the costs of transport. For this purpose, the level of geographical aggregation is much coarser (standard regions). Geographical differences in the incidence of transport costs are small and appear to be related less to the location of the region in the country and more to the density of activity within the region and the absolute size of population.

One reason for the lack of spatial variation in the cost of transport, despite

124

differences in mean haul, lies in the structure of freight charges. The limited evidence available is examined in chapter 9 and it is clear that, at least for road transport, terminal charges in fact account for a very high proportion of total charges. Even for zone 68, with Aberdeen as the centroid, terminal charges exceed half the total.

Interpretation

Perhaps the first point to note is that rail traffic in general has proved harder to model than road freight, except when linear programming is used. The main reason is that rail traffic is confined to a limited range of bulky commodities; the rail system is no longer the main carrier of general merchandise. It follows that the pattern of rail traffic is determined very largely by the locations of a limited number of producers and consumers – coal mines and electricity generating stations, steel plants and steel stockists, etc. Consequently, changes in the pattern of freight movements by rail are closely determined by a limited number of decisions, over some of which government has considerable control. More general decisions about the location of employment and population have relatively little impact on rail freight movements. On the other hand, it is precisely these sorts of decision – about the location of New Towns, regional centres of growth, etc. – which do have a big impact on road freight patterns. In terms of policy conclusions, therefore, it is with reference to road transport that most can be said.

However, there is an abiding problem on which no evidence can be offered. Our study is based on real-world data, with which two classes of model have been compared: the stochastic gravity model and the normative linear programming approach. If either of these techniques is used to predict the pattern of traffic flows, given some new set of spatially located demands and supplies, is it valid to suppose that incremental activities in each zone will conform to the pre-existing patterns of spatial interaction? In other words, we have been dealing with the average situation and there is no direct evidence to show whether the marginal activities will behave in the average manner. The difficulty is further compounded by the question of time period. Even if the marginal activity initially has a spatial interaction pattern differing from the average for the zone, it may in time converge towards that average. Such a process would be represented in the case of an immigrant firm by the progressive substitution of local suppliers for its traditional, but now more distant, suppliers. The difficulty arising from the distinction between average and marginal conditions applies particularly to questions concerning where in the country to locate employment and population. In such a case, the marginal increment may double or treble the existing population. In general, the marginal/average dilemma is less acute in modelling transport flows to assess the need for investment in routes, because marginal increments in the majority of centres are relatively small. Only in the case of a locally concentrated development might major problems arise. However, as the L.P. technique performs well for the twenty-four major urban areas in England and Wales, this is not too serious a matter, since it means that all cities over approximately 0.5 million inhabitants, or even perhaps 0.25 million, have characteristics that for our purpose are identical. At least at this level, marginal may reasonably be equated with average conditions.

If we consider first the modelling of freight flows at the level of aggregation employed, it appears that the normative view of minimising cost and effort is nearer the truth than is the stochastic concept of behaviour. If further work substantiates this conclusion, a powerful tool becomes available for predicting patterns of road traffic given future spatial arrangements of activities. The greatest weakness in such predictions lies in the estimation of freight volumes that will be generated by and attracted to zones. Nevertheless, sufficient progress has been made for it now to be possible to engage in simulation exercises to assess the likely maximum and minimum flows between pairs of zones and some 'most probable' middle value. In this way, a reasonable estimate could be obtained for the range of possible outcomes of policies regarding the location of large increments of population, as envisaged in the Humberside and Severnside studies. Such exercises would require very considerable computing capacity and back-up facilities, in excess of those available to us.

As for the question of where in the country increments of population should be housed, some fairly clear conclusions do emerge. It will be recalled that potential accessibility, measured by population-miles, does account for a reasonable proportion of the variance in both the gravity model β values and mean haul. Furthermore, freight does appear to move according to normative principles of minimising the length of haul, or selecting origins and destinations that are nearer rather than further away. In principle, therefore, the remoter locations are at a serious disadvantage in comparison with the more central ones. Note, however, tnat in relating the traffic flow characteristics to population-miles, it was the logarithm of the latter that was used. Reference to table 6.9 shows that a ten-fold increase in the value for population-miles will, other things being equal, cause the mean haul on both inwards and out-wards traffic to be some twenty miles less. As a consequence, it is really only the most remote locations, such as west Cornwall, west Wales and northern Scotland, that are at a serious disadvantage in this respect. For zones near to a major urban centre — Glasgow, Tyneside, Bristol, etc. — there is no really serious effect on mean haul owing to location within the country.

This conclusion relates to the aggregate volume of freight and is essentially only relevant, therefore, for government, taking account of the permanent consumption of real resources involved in major locational decisions. It is not a conclusion that is necessarily valid for firms or whole industries, for whom it is the private costs and benefits which are of concern. However, evidence from the 1963 census of manufacturing supports the proposition that intra-regional spatial allocation of resources has a more significant impact on the level of transport costs than does inter-regional location.

One reason for this lies in the variable length of haul for commodities and their relative importance in traffic volumes. As shown elsewhere (Chisholm, 1971a), a very large part of the tonnage of freight is essentially local in character, comprising building materials, beverages and manufactured foodstuffs, among other items. Many of these goods — sand and gravel for example — are fairly ubiquitous in their distribution and can, therefore, be supplied from local sources. Others, notably mineral waters, are distributed from widely scattered manufacturing plants, because transport costs quickly offset possible scale economies of manufacture. For this kind of commodity, spatial variations in the amount of transport work to be done are very small indeed and, as with sand and gravel, may even militate against the more central zones. At the other extreme, the volume of commodities that is distributed over a national market is in fact very small, as indicated in table 6.3. When we add to these propositions the fact that for road transport terminal charges are a very high proportion of total road freight charges, it becomes abundantly clear why comparatively small regional differences in the incidence of transport costs are all that can be detected.

From this reasoning we can also reconcile an apparent discrepancy with which we started. Location theory lays considerable stress on transport costs as a locating factor for industry. One of the major weaknesses of location theory as a basis on which to establish policy conclusions is that it gives no guidance as to the proportion of plants or firms for which scale economies in production are exhausted, or offset by transport costs, at various shares of the national market. Bain (1956), Pratten & Dean (1965) and Pratten (1971) have compiled estimates for the United States and the United Kingdom showing just how variable a proportion of the national market is accounted for by an optimum-sized plant in the manufacturing sector. The work reported here does not throw light directly on this matter but has yielded evidence that can most plausibly be interpreted to suggest that the greater part of freight traffic originates from organisations that command a very small share of the national market. If this is in fact the case, then clearly there are sharp limits to the extent to which cumulative localisation in the central areas need occur. Conversely, policies designed to maintain and reinforce the major urban centres in peripheral locations are not likely to impose a significant extra long-term cost on the nation on account of greater aggregate transport costs, so long as investment is concentrated in or near major cities.

Data

Perhaps the single most serious weakness of the freight flow data available to us is the manner in which the mean haul is estimated. The raw returns have been allocated to an origin and destination and the mean haul has then been obtained in two ways: as the distance between centroids, for inter-zonal flows; and as a function of the zone's area, in the case of intra-zonal movements. Of the two procedures, it is undoubtedly the latter which is most open to question. Now it is possible to avoid these dubious shifts of estimation if the initial tabulation of the raw returns assigns not only the tonnage but the ton-mileage of each freight movement to the zones of origin and destination. From the aggregates so obtained for each commodity, or for all commodities combined, the mean length of haul for each cell in the origin-destination matrix could be obtained by dividing the ton-mileage by the tonnage. In this way, the mean haul on all flows, both inter- and intra-zonal, would be estimated independently of the network of reporting units and assumed centroids of origin/destination.

That this suggested remedy for one major defect in the data is not unreasonable is attested by the following. At an early stage in the research here reported, negotiations were entered into with the then Ministry of Transport and their computer agency for some special tabulations of the 1967–8 road freight survey, including precisely the kind outlined in the previous paragraph. These requests were agreed in principle but it was not until mid-1972 that some data in this form became available.

The second major weakness of the data we have used is the highly variable shape and size of the traffic zones by which data are recorded. While it is true that matrices of 78 x 78 are laborious to handle, there does seem to be a clear case for at least one survey to be undertaken with a substantially greater degree of spatial disaggregation. This would permit deliberate experimentation with varying degrees of spatial aggregation, such experiments in turn serving to guide the level of geographical aggregation at which subsequent surveys would be undertaken.

Both the points made above are designed to improve the performance of static models and modes of analysis. The fact that encouraging results have been obtained on the basis of the admittedly rather inadequate 1964 data, and that some basic data improvements can be envisaged, leads to the observation that the most interesting results may not emerge until changes over time can be analysed. If the parameters of models prove highly stable over time, then their value in a predictive role is enhanced. If, on the other hand, the parameters are themselves changing, then the major research problem may be the forecasting of these changes. A necessary condition for making such forecasts is the possession of data for more than one moment of time. Either way, there is a clear need for further national surveys of freight movements, whether these are regarded as independent exercises or are linked to the periodic censuses of the manufacturing and distributive trades. When, or if, data become available for successive dates, it will be possible to follow up the work initiated by Suzuki (1971) and apply to the British case the techniques which he developed and applied to Japan. Given that he used fifteen regions covering all Japan and data for only three consecutive years, 1962, 1963 and 1964, his conclusion, that interregional trade coefficients are stable over time, is not surprising. The fact is that we do not know how stable and over what time period.

Future research

Should improved data of the kind envisaged in the previous section become available, then a number of things suggest themselves as needing attention: to repeat the kind of analyses here presented; to experiment with varying degrees of spatial aggregation; to examine the stability or otherwise of parameters over time. However, within that kind of framework, two matters stand out as urgently needing further work. The first is to improve the basis on which the volume of freight demanded in and supplied by zones can be predicted from exogenous variables, such as employment. In particular, it would seem necessary to devise different methods for rail and road traffic, given the lack of success we have had with the former. The second main need is to extend the linear programming approach to see how many of the urban areas can be efficiently included. Once an efficient means has been devised for estimating the main freight flows, it should then be possible to take this basic framework as given and experiment with means for allocating the lesser flows — possibly in terms of a gravity model based on the identified main nodes.

Gordon (1971) has suggested that the geographical variation in β values is due to the fact that no attraction constraint factor, like B_j in equation 6.1, has been used for the destinations. He noted that distance exponents tend to be higher in the less accessible zones and that therefore in calibrating a national gravity model some surrogate for the attraction factor might be incorporated, such as the population-miles measure of potential accessibility. The surmise is that if some such measure were incorporated, the β values for the zones would display less variation. In the doubly constrained gravity model for flows between all regions, the single distance parameter is assumed to be valid because of the operation of constraint factors A_i and B_j. However, this assumption, taken in conjunction with Gordon's surmise, leads us to take more seriously the objections to the gravity model raised in chapter 6.

By whatever means the gravity model is derived, it operates by averaging the distance-decay effects over all regions. The misallocations of originating and terminating traffic which arise from this parsimonious generalisation of behaviour are compensated for by the constraint factors. These are calculated by an iterative procedure which adjusts the cell entries so that the marginal totals are respected. This implies a considerable departure from the empirical/behavioural aspect of the model in favour of the requirements of the manipulation of the balancing devices. The fact that the predictions of interregional freight flows produced by constrained models are not very sensitive to variations in the β values used suggests that the constraint factors are dominant. This may arise because of the large variations in the values of the marginal totals. But whatever the cause, the fact is that one does not know the true nature of the interactions between various parts of the model. This is clearly an undesirable feature, since the empirical input is the only real-world element in the model — the rest is a book-keeping process of balancing the accounts.

Thus to return to Gordon's comment. He has suggested that for single-region gravity-model equations, experiments should be conducted to see if the spatial variation in the distance exponents would be reduced by the incorporation of some measure of accessibility of the destinations. If such a reduction could be achieved, this might validate the use of a single β value for a model using the whole matrix of flows. Such experiments should indeed be most useful, as they are likely to reveal more about the nature of the balance of power among the components of the constrained gravity model. Experiments are also needed with whole matrix models to examine the effects of different configurations of the marginal totals on the performance of the model.

Perhaps the most important need, however, is to link freight flow data more directly with producing and consuming units. This would be moving toward the input-output kind of formulation discussed in chapter 2. The key point about such an approach is the possibility that would then arise of obtaining better estimates of freight demands and supplies and providing a better interpretation of the observed pattern of freight movement. On the last point, it would then become feasible to link freight flows to economies of scale in manufacturing and processing, so as to examine further the interpretation of local advantages (or the lack thereof).

One gap in the field of freight flow modelling which urgently requires attention concerns the statistical properties of flow matrices, procedures for the estimation of parameters and the comparison of actual flows and different model outputs. The present study is undoubtedly the poorer for some inconsistencies along these lines. The difficulty is that nobody has fully examined the properties of different procedures, established the relationships between them and suggested a consistent code of practice. It is possible to correlate actual and estimated cell values for flow matrices (either including or excluding zero values); to sum squared differences between actual and estimated flows; to correlate actual and estimated trip-length frequency distributions; to carry out 'root mean square error' tests on such distributions expressed in discrete form, or to compare average lengths of haul. For parameter estimation there are various regression or iterative procedures, the relations between which are unclear. The choice of procedure to some extent depends on the objective of the exercise. For some purposes, one might wish to concentrate attention on major flows in the system, while in other cases the mean length of haul might be important. Happily, the examination of some of these problems is in hand (Gordon, 1971). In mitigation of any inconsistencies that there may be in chapter 6, we can perhaps offer the excuse that experiments on the distribution of traffic necessarily proceeded by trial and error and the available resources did not permit a full re-working of earlier trials in the light of later exercises.

Finally, an area which we have not touched on at all is the relationship between freight volume and the number of vehicle trips, an important consideration in the design of transport facilities. Of course, regulations about permissible loads and speeds etc. do change and freight transport is only part of the total system — passenger traffic being the other main element.

Enough has been said, however, to indicate beyond all reasonable doubt that a great deal still remains to be done. The present book can, therefore, be regarded as no more than a modest beginning, in which some of the ground has been explored.

Summary data for the Ministry of Transport zones

Number	Centroid	Population (000)	Employed persons (000)	Population-miles (000) $D^{1.0}$	$D^{2.5}$	Tons of freight (00) Road O_i	D_j	Total O_i	D_j	Ton-miles of freight (millions) Road O_i	D_j	Total O_i	D_j	Road β values O_i	Mean haul (miles) Road O_i	D_j	Total O_i	D_j	Standardised road mean haul (miles) O_i	Number
1	Morpeth	240	93.27	356	1.6	71,488	56,137	122,320	95,696	228.9	180.3	368.3	294.4	-2.0	32	32	30	31	29	1
2	Tynemouth	308	78.13	442	8.8	52,362	57,207	69,232	90,124	196.4	210.3	231.1	257.8	-1.6	37	36	33	29	33	2
3	Gateshead	441	197.38	624	64.0	112,427	122,850	159,063	155,340	358.6	385.2	494.9	454.0	-1.8	31	31	31	29	33	3
4	Newcastle-on-Tyne	269	172.58	733	128.3	59,265	64,756	59,782	70,818	186.8	201.1	188.1	245.5	-1.7	31	31	31	35	33	4
5	Sunderland	190	87.84	598	192.8	41,724	45,594	42,298	50,380	152.8	164.1	155.5	182.2	-2.7	36	35	37	36	33	5
6	Durham	620	225.97	438	3.0	181,692	175,402	344,779	206,207	450.2	398.7	1,138.4	470.1	-2.2	24	22	33	23	30	6
7	Darlington	84	42.86	517	86.1	24,541	23,683	25,874	27,931	65.1	58.7	77.2	72.6	-2.9	26	24	30	26	30	7
8	Stockton	299	155.06	507	25.0	95,377	93,172	153,336	240,665	260.0	214.9	646.0	802.6	-1.8	27	23	42	33	34	8
9	Middlesbrough	157	59.24	595	167.1	50,011	48,857	54,126	62,486	135.5	111.3	163.5	137.2	-2.7	27	22	30	22	34	9
10	Northallerton	304	114.92	423	1.0	76,786	71,944	82,915	84,018	327.6	287.9	348.9	350.9	-2.7	42	40	42	42	32	10
11	Selby	316	140.72	511	1.8	79,529	81,283	102,610	102,911	313.7	315.3	491.7	410.5	-2.7	39	38	48	40	32	11
12	York	104	62.79	586	105.6	26,212	26,803	27,932	28,437	101.6	100.8	124.4	119.8	-3.5	38	37	45	42	32	12
13	Leeds	1,755	878.98	724	11.1	529,236	537,222	666,610	628,852	1,253.6	1,285.3	1,971.9	1,518.1	-2.0	23	23	30	24	33	13
14	Barnsley	661	243.49	638	5.2	194,341	185,167	338,664	202,195	560.6	498.0	1,546.9	568.1	-2.3	28	26	46	28	29	14
15	Doncaster	86	54.28	666	88.6	25,265	24,064	38,884	25,483	76.7	68.1	129.5	75.9	-2.1	30	21	33	26	29	15
16	Sheffield	718	367.49	663	10.1	262,181	273,471	311,384	338,053	525.9	584.3	818.3	900.1	-1.6	20	21	26	27	37	16
17	Hull	527	227.73	407	1.3	192,942	182,762	215,250	233,783	695.1	567.3	816.5	883.6	-1.8	36	31	38	38	33	17
18	Workington	225	94.03	289	0.4	71,868	69,859	99,741	94,703	348.3	322.6	574.3	459.5	-2.8	48	46	58	49	33	18
19	Carlisle	71	34.37	396	71.4	22,701	22,056	23,142	25,832	104.1	99.5	109.1	136.4	-3.1	45	45	47	53	33	19
20	Kendal	68	31.22	400	1.1	31,426	32,378	37,461	34,404	120.6	119.2	189.3	134.9	-3.0	38	36	51	39	32	20
21	Carnforth	420	179.55	453	3.4	99,412	109,097	105,735	132,288	277.2	324.4	329.9	585.8	-2.5	27	29	31	44	32	21
22	Lancaster	90	38.69	526	21.4	21,970	24,114	27,992	34,144	53.7	59.1	128.9	141.2	-1.9	24	24	46	41	32	22
23	Leyland	166	71.93	621	7.9	62,667	67,555	62,947	71,891	172.1	182.0	174.6	201.5	-2.2	27	26	28	28	33	23
24	Preston	113	68.32	707	118.8	42,652	45,989	46,533	57,736	110.7	118.4	121.9	182.6	-1.9	25	25	26	32	33	24
25	Newton-le-Willows	282	121.73	728	11.2	105,427	100,387	128,083	125,924	242.4	220.8	271.1	329.6	-2.1	22	22	21	26	32	25
26	Wigan	79	41.20	821	89.9	27,646	26,334	34,976	27,717	63.7	58.9	77.1	68.6	-1.8	23	22	22	25	32	26
27	Warrington	76	47.37	756	82.9	26,658	25,378	29,942	40,080	63.2	57.5	82.8	99.0	-1.8	24	23	28	33	32	27
28	St Helens	108	53.45	803	115.8	37,705	35,909	51,566	51,253	92.6	84.9	118.0	125.7	-1.8	25	23	23	25	32	28
29	Accrington	554	240.71	578	6.4	145,830	138,968	152,137	156,705	437.1	381.3	468.1	582.6	-2.0	30	27	31	37	35	29
30	Bolton	810	391.95	851	21.1	216,784	236,390	218,519	249,745	617.8	697.1	630.9	780.0	-1.4	28	29	29	31	36	30
31	Manchester	716	462.11	984	59.1	212,309	231,525	221,450	256,399	531.9	598.2	592.2	786.9	-1.5	25	25	27	31	36	31
32	Stockport	833	362.16	854	25.1	210,376	229,427	211,605	244,562	574.8	637.3	587.0	719.4	-1.6	27	27	28	36	36	32
33	Ormskirk	601	228.58	675	10.8	202,739	198,531	214,753	221,154	665.7	627.3	772.7	794.0	-1.5	32	31	36	29	35	33
34	Liverpool	747	384.31	998	123.5	254,628	249,325	266,850	285,487	732.9	678.6	815.6	999.6	-1.5	28	27	31	36	35	34
35	Birkenhead	408	154.03	1,003	147.4	138,935	136,054	173,051	145,133	407.4	376.6	621.9	444.8	-1.5	29	27	36	31	35	35
36	Crewe	497	227.35	587	2.6	221,582	200,517	269,504	246,523	768.0	630.7	1,045.2	835.2	-2.8	34	31	39	34	33	36
37	Derby	1,159	523.67	678	3.9	365,613	307,460	615,957	435,365	1,375.3	965.6	3,065.4	1,333.6	-2.3	37	31	50	31	30	37
38	Nottingham	556	281.40	674	11.8	133,086	135,128	214,142	154,628	409.5	464.0	871.3	526.6	-2.4	30	34	41	34	31	38
39	Scunthorpe	430	187.05	494	1.7	185,397	151,500	225,627	276,312	825.5	541.4	1,165.6	1,287.4	-1.9	44	35	52	47	34	39
40	Northampton	1,039	497.99	664	2.7	279,272	263,449	424,969	321,089	1,204.1	1,030.2	2,214.6	1,286.5	-1.9	43	39	52	40	32	40

(contd.)

Number	Centroid	Population (000)	Employed persons (000)	Population-miles (000) $D^{1.0}$	Population-miles (000) $D^{2.5}$	Tons of freight (00) Road O_i	Tons of freight (00) Road D_j	Tons of freight (00) Total O_i	Tons of freight (00) Total D_j	Ton-miles of freight (millions) Road O_i	Ton-miles of freight (millions) Road D_j	Ton-miles of freight (millions) Total O_i	Ton-miles of freight (millions) Total D_j	Road β values β_0 O_i	Mean haul (miles) Road O_i	Mean haul (miles) Road D_j	Mean haul (miles) Total O_i	Mean haul (miles) Total D_j	Standardised road mean haul (miles) O_i	Number
41	Leicester	380	215.01	647	8.2	84,886	97,953	85,774	104,645	291.2	296.9	298.1	334.9	−2.8	34	30	35	32	33	41
42	Peterborough	74	46.02	520	3.1	49,961	43,030	50,621	44,671	245.0	166.9	262.4	180.6	−2.0	49	38	52	40	33	42
43	Huntingdon	80	39.29	534	1.5	43,085	27,675	49,510	29,656	181.5	102.3	241.9	117.4	−3.1	42	36	49	40	32	43
44	Bedford	174	77.31	603	2.6	61,530	43,373	62,310	53,362	281.2	203.8	321.0	326.4	−3.0	48	46	52	61	31	44
45	Norwich	1,033	597.54	344	0.5	420,044	437,464	437,281	468,713	1,893.8	2,064.9	2,124.4	2,464.0	−3.8	45	47	49	53	33	45
46	Stevenage	786	371.85	698	4.8	153,363	173,125	156,760	191,507	484.9	546.4	534.7	746.0	−2.6	31	31	34	39	34	46
47	Chelmsford	1,115	469.93	619	3.0	334,564	336,544	362,568	351,482	970.2	1,049.7	1,275.0	1,252.7	−2.7	28	31	35	36	33	47
48	Holborn	1,295	1,771.50	3,277	2,007.6	477,453	448,321	480,217	451,191	1,285.3	1,461.2	1,313.5	1,507.2	−1.3	26	32	27	33	36	48
49	Charing Cross	1,900	642.22	2,862	1,437.6	326,343	306,428	337,811	340,844	871.4	993.5	996.8	1,499.3	−1.3	26	32	30	44	36	49
50	Leyton	1,773	439.37	1,427	74.9	213,922	198,206	221,607	210,860	551.2	433.1	638.4	583.9	−1.9	25	21	29	28	35	50
51	Harringay	1,417	514.00	1,819	224.0	13,173	19,309	14,957	38,227	32.0	48.2	62.1	302.6	−1.6	24	24	42	79	34	51
52	Southall	2,126	398.66	1,194	34.2	333,169	349,503	333,854	358,230	774.0	862.5	785.0	963.6	−2.3	23	25	24	27	35	52
53	Croydon	954	384.63	1,237	46.4	88,614	121,582	89,044	129,164	203.8	307.4	209.4	375.1	−2.4	23	25	24	29	35	53
54	Bromley	504	149.86	1,158	36.3	45,985	61,371	46,206	63,948	121.2	138.5	122.6	168.4	−1.7	26	22	27	26	35	54
55	Crawley	3,150	1,307.60	639	3.0	588,137	517,531	635,292	666,462	1,970.2	2,069.4	2,368.4	2,520.1	−2.6	33	33	37	38	32	55
56	Southampton	1,518	700.43	474	1.9	299,762	310,267	315,151	328,141	1,023.1	1,191.2	1,152.2	1,403.8	−3.2	34	38	37	43	32	56
57	Oxford	1,294	649.54	584	1.7	296,108	300,927	312,766	323,644	1,265.7	1,213.0	1,450.2	1,415.4	−3.0	42	40	46	44	32	57
58	Bristol	2,243	1,046.40	431	0.8	578,856	577,765	603,789	614,783	2,662.2	2,648.3	2,885.1	3,114.5	−3.8	45	45	48	51	33	58
59	Plymouth	1,165	484.01	260	0.5	304,499	318,695	317,373	341,254	989.7	1,242.2	1,106.1	1,562.3	−3.0	38	38	35	46	31	59
60	Wolverhampton	1,469	708.11	577	1.5	537,818	534,795	586,867	637,178	2,265.9	2,326.4	2,506.1	2,944.8	−2.6	42	43	43	46	32	60
61	Coventry	409	229.98	768	28.9	82,844	99,079	90,215	114,311	250.6	330.8	277.6	458.0	−1.7	30	33	31	40	36	61
62	Birmingham	2,337	1,256.30	873	19.4	733,007	771,447	742,822	848,308	1,788.4	2,008.4	1,872.9	2,619.6	−1.5	24	26	25	31	37	62
63	Stoke-on-Trent	376	193.66	731	26.1	116,713	99,484	155,794	120,336	262.0	208.2	400.7	326.7	−2.3	22	20	26	27	31	63
64	Luton	211	118.89	722	9.7	72,888	58,342	82,442	58,342	272.8	163.9	369.5	185.9	−2.1	37	30	45	33	33	64
65	Cardiff	1,704	718.26	410	2.1	473,827	459,856	806,411	787,584	1,721.9	1,565.0	3,075.2	2,841.2	−2.9	36	34	38	36	33	65
66	Aberystwyth	442	173.04	310	0.3	149,321	158,508	193,996	186,967	671.8	773.2	1,097.1	983.9	−3.2	44	48	57	53	30	66
67	Rhyl	496	196.26	407	1.1	152,226	132,875	175,140	186,209	550.0	439.4	678.1	759.0	−2.8	36	33	39	41	31	67
68	Aberdeen	959	385.70	158	0.1	345,665	349,807	349,490	185,715	2,398.9	2,389.4	2,495.4	2,845.8	−4.8	69	68	71	74	32	68
69	Dundee	553	243.93	222	0.7	160,671	159,702	224,548	193,950	582.0	559.7	1,101.4	765.8	−3.6	36	35	49	39	32	69
70	Falkirk	261	82.09	319	1.8	145,916	104,940	160,395	121,621	427.7	355.9	540.3	451.9	−3.2	29	31	34	37	33	70
71	Paisley	431	232.39	487	22.0	121,008	138,076	121,207	142,537	300.8	398.4	303.1	420.7	−2.7	28	28	25	30	34	71
72	Glasgow Central	1,055	472.69	667	72.0	203,222	231,875	237,153	261,966	421.6	564.2	536.0	785.9	−2.4	24	23	23	30	34	72
73	Motherwell	329	143.44	436	13.4	159,662	182,187	198,930	271,313	385.0	489.1	538.3	971.4	−2.5	26	27	27	36	34	73
74	Ayr	363	148.89	272	1.0	109,173	110,718	141,195	133,354	299.5	304.6	444.8	409.8	−3.2	27	27	31	31	31	74
75	Lanark	335	119.71	333	1.8	119,272	108,169	147,087	123,559	289.8	272.9	370.6	349.2	−3.1	25	25	25	28	29	75
76	Edinburgh	542	268.70	367	10.1	140,968	127,731	174,727	149,870	405.8	354.8	506.0	425.5	−3.3	27	28	29	28	32	76
77	Hawick	190	77.46	285	0.4	44,871	44,292	48,991	47,569	187.3	184.6	234.5	206.7	−2.3	41	41	48	43	31	77
78	Dumfries	147	60.17	282	0.3	61,819	67,650	72,498	80,214	224.3	255.7	358.8	417.3	−2.7	36	37	49	52	31	78
Total		52,117	24,060.74			13,836,407		16,294,140		45,891.6		60,592.1								
Mean				678.5	76.2									−2.39	32.09	31.46	36.73	36.69	32.83	
Standard deviation				479.5	277.2									0.689	8.72	8.26	10.33	10.53	1.89	

For discussion of the data sources, etc., see chapter 4 and pp. 37–40.
Derivation of the standardized road mean haul is discussed on p. 98.

Outline identification map of traffic zones.

LIST OF REFERENCES

Bain, J. S. (1956). *Barriers to New Competition,* Harvard University Press.

Bayliss, B. T. & Edwards, S. L. (1970). *Industrial Demand for Transport,* H.M.S.O.

Black, W. R. (1971). 'The utility of the gravity model and estimates of its parameters in commodity flow studies', *Proceedings of the Association of American Geographers,* **3,** 28–32.

Board of Trade (1970). *Report on the Census of Production 1963,* H.M.S.O.

Bowers, J. (1970). 'The anatomy of regional activity rates', *Regional Papers,* **1,** National Institute of Economic and Social Research, Cambridge University Press.

British Railways Board (1965). *The Development of the Major Railway Trunk Routes,* British Railways Board.

Britton, J. N. H. (1967). *Regional Analysis and Economic Geography,* Bell.

Brown, A. J. (1969). 'Surveys of applied economics: regional economics, with special reference to the United Kingdom', *Economic Journal,* 79, 759–96.

Central Unit for Environmental Planning (1971). *Severnside: a feasibility study,* H.M.S.O.

Chenery, H., Clark, P. G. & Cao-Pinna, V. (1953). *The Structure and Growth of the Italian Economy,* U.S. Mutual Security Agency, Rome.

Chisholm, M. (1971a). 'Freight transport costs, industrial location and regional development', in M. Chisholm and G. Manners (eds.) *Spatial Policy Problems of the British Economy,* Cambridge University Press, 213–44.

Chisholm, M. (1971b). 'In search of a basis for location theory: micro-economics or welfare economics?', in C. Board *et al.* (eds.) *Progress in Geography,* vol. 3, Arnold, 111–33.

Chisholm, M. (1971c). 'Forecasting the generation of freight traffic in Great Britain', in M. Chisholm, A. E. Frey and P. Haggett (eds.) *Regional Forecasting,* Butterworths.

Clark, C. (1966). 'Industrial location and economic potential', *Lloyds Bank Review,* 82, 1–17.

Clark, C., Wilson, F. & Bradley, J. (1969). 'Industrial location and economic potential in Western Europe', *Regional Studies,* 3, 197–212.

Deakin, B. M. & Seward, T. (1969). 'Productivity in Transport. A study of employment, capital, output, productivity and technical change', University of Cambridge, Department of Applied Economics, *Occasional Papers,* no. 17. Cambridge University Press.

Deakin, B. M. & Seward, T. (1970). Private communication.

Department of Economic Affairs (1969). *Progress Report,* 53.

Department of the Environment (1971). *Initial Attempts at Modelling Road Freight Flows,* Mathematical Advisory Unit, note 211 (mimeo).

Dorfman, R., Samuelson, P. A. & Solow, R. M. (1958). *Linear Programming and Economic Analysis,* McGraw-Hill.

Dunford, M. (1969). 'Interregional Freight Flows: an application of linear programming', unpublished B.Sc. dissertation, Department of Geography, University of Bristol.

Edwards, S. L. (1967). *The West Cumberland Transport Survey,* Northern Economic Planning Board.

Edwards, S. L. (1970a). 'Transport cost in British industry', *Journal of Transport Economics and Policy,* 4, 1–19.

Edwards, S. L. (1970b). Private communication.

Freight flows and the British economy

Edwards, S. L. & Gordon, I. R. (1971). 'The application of input-output methods to regional forecasting: the British experience', in M. Chisholm, A. E. Frey and P. Haggett (eds.) *Regional* Butterworths, 415–30.

Frost, M. (1969). 'Distribution costs as a factor in the location of industry policy', *Discussion Paper* 34, Department of Geography, London School of Economics and Political Science.

Gordon, I. R. (1971). Private communication.

Haggett, P. & Chorley, R. J. (1969). *Network Analysis in Geography,* Arnold.

Harris, C. D. (1954). 'The market as a factor in the localization of industry in the United States', *Annals,* Association of American Geographers, 44, 315–31. Reprinted in R. H. T. Smith, E. J. Taaffe and L. J. King, *Readings in Economic Geography. The location of economic activity*, Rand McNally (1968) 186–99.

Heady, E. O. & Skold, M. D. (1966). 'Analyses to specify the regional distribution of farm products', in *Research and Education for Regional and Area Development,* Iowa University Press, 175–92.

Henderson, J. M. (1958). *The Efficiency of the Coal Industry: an application of linear programming,* Harvard University Press.

Herington, E. L. (1971). 'A Probabilistic Adaptation of Linear Programming', unpublished B.Sc. dissertation, Department of Geography, University of Bristol.

Ingram, D. R. (1971). 'The concept of accessibility: a search for an operational form', *Regional Studies,* 5, 101–7.

Isard, W. (1956). *Location and Space-Economy. A general theory relating to industrial location, market areas, land use, trade, and urban structure,* Wiley.

Isard, W. *et al.* (1969). *General Theory. Social, political, economic and regional,* M.I.T. Press.

Kirby, H. (1969). Private communication.

Land, A. H. (1957). 'An application of linear programming to the transport of coking coal', *Journal of the Royal Statistical Society,* Ser. A, 120, 308–19.

Leontief, W. & Strout, A. (1966). 'Multiregional input-output analysis', in W. Leontief (ed.), *Input-output Economics,* Oxford University Press.

Lösch, A. (1954). *The Economics of Location,* Yale University Press. Translated from the German by W. F. Stolper.

McCrone, G. (1969). *Regional Policy in Britain,* Allen and Unwin.

Mathematica (1968). *Studies on the Demand for Freight Transportation,* 1, Northeast Corridor Transportation Project, U.S. Department of Transportation.

Mennes, L. B. M., Tinbergen, J. & Waardenburg, J. G. (1969). *The Element of Space in Development Planning,* North-Holland.

Ministry of Transport (1962). *Report of the Committee of Enquiry into the Major Ports of Great Britain,* Cmnd. 1824, H.M.S.O.

Ministry of Transport (1964). *Survey of Road Goods Transport 1962. Final results,* part I, H.M.S.O.

Ministry of Transport (1966a). *Portbury. Reasons for the Minister's decision not to authorise the construction of a new dock at Bristol,* H.M.S.O.

Ministry of Transport (1966b). *Survey of Road Goods Transport 1962. Final results, geographical analysis,* H.M.S.O.

Ministry of Transport (1966c). *Survey of Road Goods Transport 1962. Methodological report,* H.M.S.O.

Ministry of Transport (1968). *Road Goods Survey 1962: the magnetic tapes from the geographical analysis and their use in the T.C.M.,* Mathematical Advisory Unit, note 101 (mimeo).

Morrill, R. (1967). 'The movement of persons and the transportation problem', in W. Garrison and D. Marble (eds.), *Quantitative Geography,* Northwestern University Studies in Geography, no. 14.

Myrdal, G. (1957). *Economic Theory and Under-Developed Regions,* Duckworth.

National Ports Council (1966). *Digest of Port Statistics 1966,* National Ports Council.

136

Olsson, G. (1965). *Distance and Human Interaction. A review and bibliography,* Regional Science Research Institute.

O'Sullivan, P. (1968). 'An analysis of Irish interregional freight flows', *Discussion Paper* 28, Department of Geography, London School of Economics and Political Science.

O'Sullivan, P. (1969). *Transport and the Irish Economy,* London School of Economics and Political Science, Geographical Papers.

O'Sullivan, P. (1971). 'Forecasting interregional freight flows in Great Britain', in M. Chisholm, A. E. Frey and P. Haggett (eds.) *Regional Forecasting,* Butterworths, 443–50.

Peaker, A. (1971). 'Regional growth and economic potential – a dynamic analysis', *Regional Studies,* 5, 49–54.

Pratten, C. (1971). 'Economies of Scale in Manufacturing Industry', University of Cambridge, Department of Applied Economics, *Occasional Papers,* no. 28, Cambridge University Press.

Pratten, C. & Dean, R. M. (1965). 'The Economies of Large-scale Production in British Industry', University of Cambridge, Department of Applied Economics, *Occasional Papers,* no. 3, Cambridge University Press.

Prest, A. R. & Turvey, R. (1965). 'Cost-benefit analysis: a survey', *Economic Journal,* 75, 683–735.

Quandt, R. E. & Baumol, W. J. (1966). 'The demand for abstract modes: theory and measurement', *Journal of Regional Science,* 6, 13–26.

Quarmby, D. A. (1967). 'Choice of travel mode for the journey to work: some findings', *Journal of Transport Economics and Policy,* 1, 1–42.

Starkie, D. N. M. (1967). *Traffic and Industry. A study of traffic generation and spatial interaction,* London School of Economics and Political Science, Geographical Papers.

Stevens, B. H. (1961). 'Linear programming and location rent', *Journal of Regional Science,* 3, 15–26.

Stewart, J. Q. & Warntz, W. (1958). 'Physics of population distribution', *Journal of Regional Science,* 1, 99–123. Reprinted in B. J. L. Berry and D. F. Marble, *Spatial Analysis. A reader in statistical geography,* Prentice-Hall (1968), 130–46.

Suzuki, K. (1971). 'Observations on the stability of the structure of the interregional flow of goods', *Journal of Regional Science,* 11, 187–209.

United Kingdom, (1970). *Annual Abstract of Statistics,* 107, H.M.S.O.

Warner, S. L. (1962). *Stochastic Choice of Mode in Urban Travel: a study in binary choice,* The Transportation Center, Northwestern University.

Weber, A. (1929). *Theory of the Location of Industries,* University of Chicago Press. Translated from the German by C. J. Friedrich.

Wilson, A. G. (1967). 'A statistical theory of spatial distribution models', *Transportation Research,* 1, 253–69.

Wilson, A. G. (1968). 'Interregional commodity flows: entropy maximising approaches', Centre for Environmental Studies, *Working Paper* 19.

INDEX

139

Index